GEOLOGY AND MINERALOGY RESEARCH DEVELOPMENTS

HALLOYSITE

STRUCTURE, OCCURRENCE AND APPLICATIONS

GEOLOGY AND MINERALOGY RESEARCH DEVELOPMENTS

Additional books and e-books in this series can be found on Nova's website under the Series tab.

GEOLOGY AND MINERALOGY RESEARCH DEVELOPMENTS

HALLOYSITE

STRUCTURE, OCCURRENCE AND APPLICATIONS

HERBERT A. ECKART
EDITOR

Copyright © 2020 by Nova Science Publishers, Inc.

All rights reserved. No part of this book may be reproduced, stored in a retrieval system or transmitted in any form or by any means: electronic, electrostatic, magnetic, tape, mechanical photocopying, recording or otherwise without the written permission of the Publisher.

We have partnered with Copyright Clearance Center to make it easy for you to obtain permissions to reuse content from this publication. Simply navigate to this publication's page on Nova's website and locate the "Get Permission" button below the title description. This button is linked directly to the title's permission page on copyright.com. Alternatively, you can visit copyright.com and search by title, ISBN, or ISSN.

For further questions about using the service on copyright.com, please contact:
Copyright Clearance Center
Phone: +1-(978) 750-8400   Fax: +1-(978) 750-4470   E-mail: info@copyright.com.

NOTICE TO THE READER

The Publisher has taken reasonable care in the preparation of this book, but makes no expressed or implied warranty of any kind and assumes no responsibility for any errors or omissions. No liability is assumed for incidental or consequential damages in connection with or arising out of information contained in this book. The Publisher shall not be liable for any special, consequential, or exemplary damages resulting, in whole or in part, from the readers' use of, or reliance upon, this material. Any parts of this book based on government reports are so indicated and copyright is claimed for those parts to the extent applicable to compilations of such works.

Independent verification should be sought for any data, advice or recommendations contained in this book. In addition, no responsibility is assumed by the Publisher for any injury and/or damage to persons or property arising from any methods, products, instructions, ideas or otherwise contained in this publication.

This publication is designed to provide accurate and authoritative information with regard to the subject matter covered herein. It is sold with the clear understanding that the Publisher is not engaged in rendering legal or any other professional services. If legal or any other expert assistance is required, the services of a competent person should be sought. FROM A DECLARATION OF PARTICIPANTS JOINTLY ADOPTED BY A COMMITTEE OF THE AMERICAN BAR ASSOCIATION AND A COMMITTEE OF PUBLISHERS.

Additional color graphics may be available in the e-book version of this book.

Library of Congress Cataloging-in-Publication Data

ISBN: 978-1-53616-812-9

Published by Nova Science Publishers, Inc. † New York

Contents

Preface		vii
Chapter 1	A General Approach to Halloysite Clay Mineral *Meltem Karaismailoğlu, Sena Zeynep Kutlu and Tuğba Ucar Demir*	1
Chapter 2	Halloysite-Based Nanopeapod Composites: Rapid and Controlled *In-Situ* Growth of Noble Metal Nanostructures within Halloysite Clay Nanotubes *Taha Rostamzadeh, Md Shahidul Islam Khan and John B. Wiley*	17
Chapter 3	The Halloysite Nanostructure-Based Materials *Ayesha Kausar*	41
Chapter 4	Recent Advances in Halloysite Nanotubes Applications *Dorel Florea, Mihai Cosmin Corobea and Zina Vuluga*	57
Index		175
Related Nova Publications		185

PREFACE

Halloysite is a natural clay mineral that belongs to the kaolin group. Its chemical formula is Al2(OH)4Si2O5.nH2O. Primarily, this mineral is an ideal nanofiller for polymer composites, but it is also utilized in the manufacture of ceramic wares and in the encapsulation of drugs in the pharmaceutical industry.

The authors present details regarding a simple, rapid, high-yield, and low-temperature synthetic approach for the in-situ growth of noble metal nanoparticles in the interior of halloysite nanotubes.

Benefits of halloysite nanotubes for polymeric structures and technical applications in various fields are highlighted, particularly their thermal stability, mechanical strength, and non-flammability.

The closing chapter focuses on the latest research concerning halloysite nanotubes applications in different fields such as biomedicine and pharmacotherapy, food packaging, agriculture, water treatment, catalysis and antifouling.

Chapter 1 - Halloysite is a natural clay mineral that belongs to the kaolin group. Its chemical formula is $Al_2(OH)_4Si_2O_5.nH_2O$. When n is equal to 2, it is hydrated, and when n is equal to 0, the halloysite is dehydrated. Generally, halloysite has a tubular structure; however, as a result of different deposits and crystallization conditions, this mineral can consist of spheroidal and plate-like particles. Its main deposits are in the United States, China,

New Zealand, Mexico, Brazil, and Turkey where halloysite is formed by weathering or hydrothermal alteration of pumice, ultramafic, and volcanic rocks. Halloysite can be used in various fields due to its characteristic features. Primarily, this mineral is an ideal nanofiller for polymer composites. It is also utilized in the manufacture of ceramic wares and encapsulation of drugs in the pharmaceutical industry. Moreover, its higher specific surface area makes halloysite a good candidate for catalyst support.

Chapter 2 - Incorporation of nanoparticles (NPs)/nanomaterials into the hollow space of naturally available tubular nanoclays can lead to the creation of advanced functional nanopeapod (NPP) composites. In this chapter, the authors present details on a simple, rapid ($\leq$ 2 min), high-yield, and low-temperature synthetic approach for the *in-situ* growth of noble metal NPs in the interior of halloysite nanotubes (HNTs). Proper combination of metal precursors in mixed polar-nonpolar solvents, surfactants and reducing agents with mild heating readily leads to the *in-situ* formation of Au, Ag, and Pd nanostructures within HNTs. The sizes of NPs are tuned mainly by adjusting nucleation and growth rates. Further modification of the process, through an increase in the amount of reducing agent, allows for the formation of nanorods (NRs)/nanowires (NWs) within the HNTs. The ability to readily grow such core–shell nanocomposites advances several applications of HNTs in catalysis, nanoreactors, water treatment, as well as in drug delivery/release systems.

Chapter 3 - The halloysite nanotubes are naturally occurring materials. The halloysite nanotubes are promising candidates for manufacturing polymer-based nanocomposites. The halloysite nanotubes have large surface area and range of exciting nanofiller properties such as mechanical strength, thermal stability, crystallization, and noncytotoxicity. In this chapter, the processing and properties of various polymer/halloysite nanotube materials have been discussed. Benefits of the halloysite nanotubes for the polymeric structures and technical applications in various fields are also highlighted. Especially, the thermal stability, mechanical strength, and non-flammability of the polymeric nanocomposite have been enhanced using the halloysite addition. The future research is needed to further develop the potential of polymer/halloysite nanotube materials.

Chapter 4 - Halloysite nanotubes (HNT) were proven as versatile structures because of their outstanding properties such as tubular or "scroll-like" morphology, high mechanical strength and thermal stability. Moreover, their crystalline structure and low hydroxyl density on the surface helps them to disperse relative easily in various polymer matrices. They are considered as cost accessible and can be found in both natural and synthetic form. HNT general availability, recommended them already for large number of applications. The present chapter summarizes the latest research concerning the HNT applications, in different fields and topics like biomedicine and pharmacotherapy, food packaging, agriculture, water treatment, catalysis, or antifouling. A special attention was given to HNT as reinforcement for polymer composites (based on different polymer matrix like PA6, PA11, PMMA, PLA, PS, PP, PVA, EVA, PHB, PEG, thermoplastic starch, or CMC). This chapter covers research on the use of Halloysite for the last 4 years.

In: Halloysite
Editor: Herbert A. Eckart
ISBN: 978-1-53616-812-9
© 2020 Nova Science Publishers, Inc.

Chapter 1

A GENERAL APPROACH TO HALLOYSITE CLAY MINERAL

Meltem Karaismailoğlu[1,*], Sena Zeynep Kutlu[2] and Tuğba Ucar Demir[2]
[1]Department of Energy Science and Technology,
Turkish-German University, Istanbul, Turkey
[2]ESAN Eczacibasi Corporation, Istanbul, Turkey

ABSTRACT

Halloysite is a natural clay mineral that belongs to the kaolin group. Its chemical formula is $Al_2(OH)_4Si_2O_5 \cdot nH_2O$. When n is equal to 2, it is hydrated, and when n is equal to 0, the halloysite is dehydrated. Generally, halloysite has a tubular structure; however, as a result of different deposits and crystallization conditions, this mineral can consist of spheroidal and plate-like particles. Its main deposits are in the United States, China, New Zealand, Mexico, Brazil, and Turkey where halloysite is formed by weathering or hydrothermal alteration of pumice, ultramafic, and volcanic rocks. Halloysite can be used in various fields due to its characteristic features. Primarily, this mineral is an ideal nanofiller for polymer

---

[*] Corresponding Author's E-mail: meltemkoglu@hotmail.com.

composites. It is also utilized in the manufacture of ceramic wares and encapsulation of drugs in the pharmaceutical industry. Moreover, its higher specific surface area makes halloysite a good candidate for catalyst support.

Keywords: halloysite, kaolinite, clay mineral, occurence

INTRODUCTION

Nanotubes have unique physical and chemical properties. They offer significant advantages for the electrical, optical, catalytic and biological applications (Ding et al. 2008; Chen et al. 2005; Thorpe et al. 2002). Halloysite is a nanoscale mineral and attracts attention because of its chemical and physical properties, occurrence and applications. The mineral belongs to the phyllosilicate group and is mainly associated with other minerals such as kaolinite, zeolite, and quartz (Ross and Kerr 1934; Margenot et al. 2017). Halloysite is naturally generated and occurs widely in weathered rocks and soils. It can be formed as various types in varied crystallization conditions and geological environments. This clay mineral is available in high quantity around the world, and its low production cost attracts attention for industrial applications. Halloysite is mainly used in the production of porcelain and crucibles (Joussein 2016). In addition to application in ceramics, halloysite can be utilized in the polymeric industry (Abdullayev et al. 2013). Its nanotubular structure makes halloysite a potential alternative to carbon nanotubes (Sudarshan et al. 2007).

This chapter documents the structure and occurrence of halloysite and discusses its potential applications with respect to its physical and chemical properties. Turkey possesses halloysite rich deposits. The halloysite sample from Balıkesir, Turkey was characteristically analyzed, and the results are presented in this chapter for the confirmation of its general structure by our own analyses.

STRUCTURAL PROPERTIES OF HALLOYSITE MINERAL

As a clay mineral, halloysite belongs to the kaolin group but is distinguished from kaolinite by its higher water content. It has two forms: hydrated and dehydrated. Determining its chemical composition is difficult because of the impurities in the mineral, but its chemical formula is $Al_2(OH)_4Si_2O_5 \cdot nH_2O$ (White et al. 2002). Halloysite-(10 Å) is the hydrated form when n is equal to 2 (Joussein et al. 2005). This mineral can be irreversibly dehydrated as the monolayer of water molecules, which forms dehydrated halloysite-(7 Å) (when n = 0), can be easily removed (Joussein et al. 2005). In a mixture of kaolinite and hydrated halloysite, halloysite can be easily distinguished by its X-ray diffraction pattern. Hydrated halloysite has a reflection by 10 Å, and the intensity of its peak helps to estimate the quantity of halloysite (Churchman 2006). In Figure 1(a), the X-ray diffraction (XRD) patterns of the halloysite sample from Balıkesir, Turkey (provided by Eczacibasi Esan) are presented, and the XRD analysis indicates that this halloysite sample consists of hydrated and dehydrated forms (Levis and Deasy 1908). The characteristic peak of hydrated halloysite can be seen at the 2θ value of 8.75° (d = 10.09 Å). Further proof of hydrated state is the existence of the peaks at 24.6° (d = 3.61 Å) and 20.58° (d = 4.31 Å). The peaks at 2θ values of 11.86° and 20.14° are assigned to the presence of dehydrated halloysite phase. In addition to the halloysite crystal phase, the existence of quartz (SiO_2) phase can also be confirmed by XRD analysis. Peaks at 20.86° and 26.64° correspond to the quartz phase. However, thermal treatment leads to a change in the crystal structure of halloysite mineral (Ouyang et al. 2014). Figure 1(b) displays the XRD patterns of the halloysite sample, which was calcined at the temperature of 1000°C. After being calcined to 1000°C, phase transformation of the crystal structure and phase segregation occur. The reflections assigned to hydrated and dehydrated halloysite phases disappear, which can be explained by the elimination of water molecules and hydroxyl groups in the inner surface (Ouyang et al. 2014). Results indicate that halloysite calcined at 1000°C consists of cristobalite (SiO_2) and mullite ($Al_{4.44}O_{9.78}Si_{1.56}$) phases. The peaks, which are located at 2θ = 21.84°, 31.16°, and 35.94° belong to the cristobalite phase. Additionally, the peaks at 2θ values of 26.01°, 26.32°, and 35.29° can be assigned to the mullite phase.

Drits et al. have used an XRD modeling approach for the determination of the crystal phases in dehydrated and partially dehydrated halloysite samples, and revealed that the calculated results are in good agreement with the experimental measurements (Drits et al. 2018). The modeling XRD patterns indicate that the samples contain either single or multi-phase crystal structure.

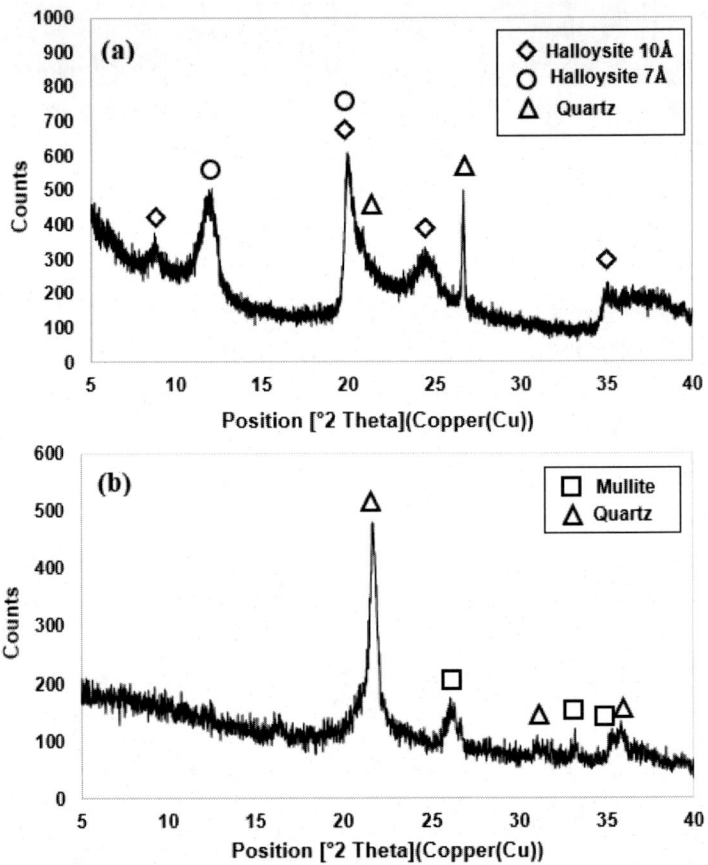

Figure 1. XRD patterns of the halloysite sample from Balıkesir, Turkey (a) and (b) halloysite sample calcined at the temperature of 1000°C.

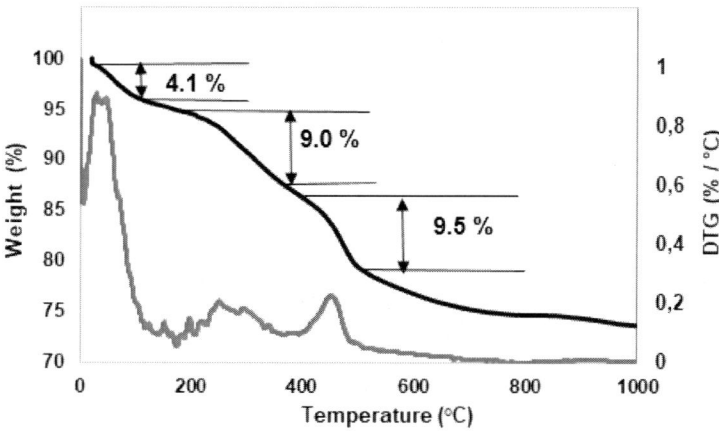

Figure 2. TGA curve of halloysite sample from Balıkesir, Turkey.

Halloysite exhibits a high specific surface area. Churchman et al. have measured the specific surface area of various halloysite samples with Brunauer, Emmett and Teller (BET) surface analysis technique, and reported that values of specific surface area are in the range of 30-170 m^2/g (Churchman et al. 1995). The halloysite sample from Balıkesir, Turkey with the specific surface area of 134.79 m^2/g exhibits relatively high specific surface area compared to the samples from other deposits (Churchman et al. 1995; Yuan et al. 2015).

Water molecules between the interlayers of halloysite can be removed even at ambient conditions (Churchman 2006), and the mineral changes to its dehydrated form. Hence, thermogravimetric analysis (TGA) is essential for the determination of the thermal behaviour of halloysite from ambient to higher temperatures. Figure 2 shows the thermal behaviour of the halloysite sample. DTG profile of the halloysite possesses an endothermic peak at around 50°C, which can be attributed to the removal of water molecules from the surface of halloysite particles (Cheng et al. 2010). In accordance with the TGA result, the elimination of water molecules causes the weight change of 4.1% around this temperature. The second endothermic peak is observed in the temperature range of 215 to 380°C, and a weight loss of 9.0% can be ascribed to the elimination of the structural water (Falcón et al. 2015). At temperatures higher than 420°C, the decomposition of halloysite takes place, and it leads to a phase segregation, which can be verified by XRD analysis (Garcia et al. 2009; Liu et al. 2013).

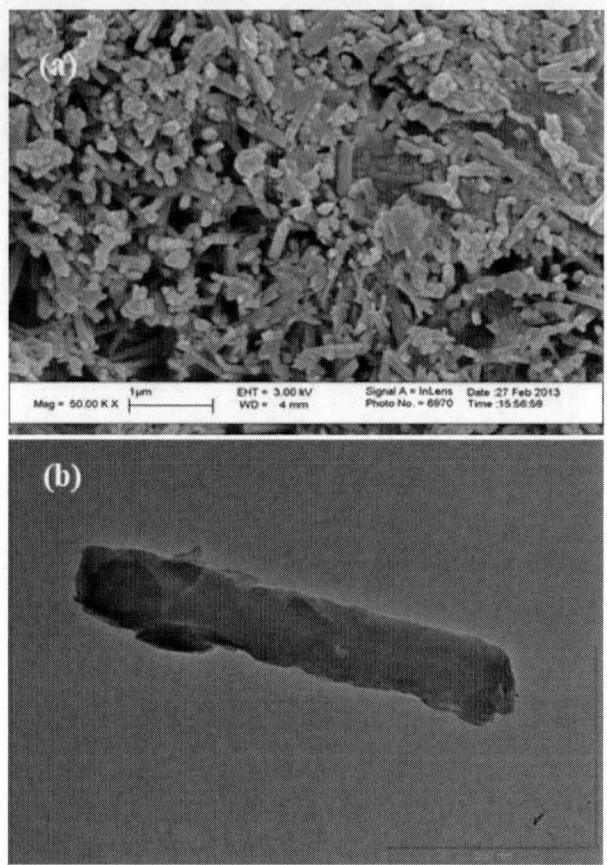

Figure 3. SEM (a) and TEM (b) micrographs of the halloysite sample from Balıkesir, Turkey.

Kaolinite and dehydrated halloysite have similar crystal phases (in terms of XRD patterns), but different morphologies (Plançon 2001). Kaolinite displays a platy-shaped particle morphology, while halloysite particles have cylindrical or tubular structure (Hart et al. 2002). However, there are other morphological forms of halloysite, such as spheroidal, platy and prismatic. Electron microscopy techniques can be used for the determination of its morphology. Figure 3(a) shows the scanning electron microscopy (SEM) micrograph of the halloysite sample from Balıkesir, Turkey at 50 000X magnification, and it can be seen that this halloysite sample exhibits a tubular morphology. In addition to the SEM micrograph, the transmitted electron microscopy (TEM) image (Figure 3(b)) displays the tubular shape of the halloysite particle.

The misfit of a larger tetrahedral sheet linked to a smaller octahedral sheet is the reason for the rolling of halloysite and the formation of tubular morphology (Bailey 1990). The surface of the tube contains silica (SiO_2) and its inner lumen consists of alumina (Al_2O_3) (Yendluri et al. 2017). Principally, a halloysite nanotube has a length of 800 ± 300 nm and its outer diameter is 15 ± 5 nm (Yendluri et al. 2017). Previous characteristic analyses proved the presence of a significant amount of iron in halloysite (De Souza Santos et al. 1966; Noro 1986), and iron can be found both in the oxide and isomorphous form. Iron content in halloysite mineral has an effect on its morphology (Hart et al. 2002). The substitution of Fe^{3+} for octahedral Al^{3+} decreases the rolling effect, and it leads to the formation of platy halloysite particles (Noro 1986; Bailey 1990). Hence, the platy form of halloysite has a large quantity of iron in comparison to its tubular form.

OCCURRENCE OF HALLOYSITE MINERAL

Temperature, time, fluid to rock ratio, and source (parent) rock are the main parameters that have an effect on the formation of halloysite mineral. It can be formed from rock types such as granite, syenite, quartz, monzonite, and silestone (Parham 1969; Wilson and Keeling 2016). Halloysite can be found in soils as well as weathered and hydrothermally altered rocks. Tubular and platy halloysite crystals are abundant in volcanic and ash soils (Hart et al. 2002). These deposits are classified into three types: deposits with high concentrations of halloysite, high-quality halloysite, and mixture of halloysite and kaolinite. High quantity of halloysite deposits are in the USA, China, and Turkey, where halloysite mineral is often precipitated through the flow of alumina-silica-rich solutions along fractures and faults of the natural rock (Wilson and Keeling 2016). The Dragon Mine in Utah

(USA) is one of the highest quality halloysite deposits, and it is also a large iron oxide source. Guizhou, Yunnan and Hunan provinces in China have small halloysite deposits (Wilson 2004). Halloysite in the Guizhou area formed mainly through alteration of volcanic rocks next to limestone and is utilized in the production of ceramics, porcelain, and bone china. There are many deposits in northwestern Turkey where the Eczacibasi Esan group extracts halloysite. (Ece and Schroeder 2007). Turplu and Tabankoy located in the Biga Peninsula contain the main deposits, in which halloysite forms in an acidic environment through alteration of andesitic tuffs (Ece et al. 2008). There are two quarries in Turplu. Here, up to 2000 tpa halloysite production can be achieved, and a small quantity of gibbsite is extracted as a byproduct. There are halloysite deposits in New Zealand, which were generated here by alteration of rhyolite (Murray 1991). The halloysite in New Zealand exhibits a similar chemical content to the halloysite from Balıkesir in Turkey (Saklar et al. 2012). Brazil is one of the producers of processed kaolin. Kaolinitic clay usually contains kaolinite and halloysite. Kaolin thus extracted in the Minas Gerais region consists of kaolinite and 7 Å/10 Å halloysite. Moreover, a small amount of iron and titania is also present in this mixture (Wilson et al. 2006). The Carlsbad and Lechugilla caves in the Guadalupe Mountains of southeastern New Mexico contain a large quantity of clay minerals. Kaolinite and hydrated halloysite have been reported as two constituents of clay minerals in these caves (Polyak and Güven 2000). Hydrated halloysite evolves in the speleogenesis by sulphuric acid (H_2SO_4) (Jagnow et al. 2000), and it can have various colours, including white, blue, and brown (Polyak and Güven 2000). This type of halloysite has a tubular form and varies in diameter and length. Generally, its diameter is less than 1.0 μm and its length is less than 0.05 μm (Polyak and Güven 2000).

APPLICATIONS OF HALLOYSITE MINERAL

Tubular halloysite exhibits significant mechanical properties and biocompatibility. Hence, it has a wide range of applications in pharmaceutical polymer and ceramic industry. Because of its low price and biocompatible nature, the use of tubular halloysite can lead to an increase in drug loading efficiency (Yendluri et al. 2017). Halloysite has a large surface area, and its internal surface is positively charged. These features enable the loading of negatively charged biomacromolecules onto the positive surface of halloysite (Fakhrullina et al. 2015). Tan et al. have investigated the performance of halloysite as a carrier. They functionalized the natural halloysite in the presence of 3-aminopropyltriethoxysilane (APTES) and loaded it with ibuprofen (Tan et al. 2013). Lun et al. have studied APTES modified halloysite nanotubes as aspirin carriers, and remarked that the APTES modified halloysite nanotubes have potential as carriers in drug industry (Lun et al. 2014). But it should be mentioned that the halloysite mineral cannot be injected, because it is non-biodegradable and may cause thrombosis (Yendluri et al. 2017). Hence, it can only be delivered orally. Halloysite can be used for the synthesis of corrosion inhibitors. In a study by Shchukin et al., a silica-zirconia-based hybrid film was doped with 2-mercaptobenzolthiazole (as a corrosion inhibitor) and loaded into a halloysite nanotube (Shchukin et al. 2008) via a sol-gel method. The results indicated that a doped hybrid film exhibited stronger corrosion protection compared to the undoped film.

Polymers are widely used for electrical, magnetic, and thermosensitive applications. Mechanical properties and thermal stability of polymers can be improved by using halloysite nanotubes in composite materials. The effect of halloysite nanotubes on the mechanical properties of linear low density polyethylene (LLDPE) has been investigated (Jia et al. 2009). Jia et al. reported that the use of halloysite nanotubes increased the tensile and flexural strength of LLDPE, and improved its flame retardancy. Ismail et al. have prepared ethylene propylene diene monomer/halloysite (EPDM/HNT) nanotube nanocomposites, and their thermal and mechanical properties, such as tensile strength, were determined (Ismail et al. 2008). The results have revealed that there is a relation between the quantity of halloysite in an EPDM/HNT nanocomposite and its mechanical properties. An increase in halloysite nanotubes in the EPDM/HNT nanocomposite led to a significant enhancement in its tensile strength, elongation at break, and tensile modulus at 100% elongation. The dispersion of halloysite nanotubes into the EPDM and inter-tubular interactions between halloysite nanotubes and EPDM are the main reasons for the increase in tensile strength. Biocompatibility of halloysite has been studied by Vergaro et al. (Vergaro et al. 2010). The halloysite nanotubes' interaction with cells was examined. Quantitative trypan blue and MTT measurements were carried out for the general cytotoxicity test. According to the results of the analysis, halloysite is a nontoxic clay mineral up to the concentration of 75 μg/mL, which makes it suitable for medical and household products. Polyvinyl alcohol (PVA) is a synthetic polymer, which exhibits excellent mechanical properties in the dry condition. However, humidity affects its mechanical properties and restricts its application. Zhou et al. have reported that the use of nanoclay such as halloysite is a convenient approach for the production of high performance PVA nanocomposites (Zhou et al. 2010). This group's research was conducted as follows. PVA and PVA/halloysite nanotubes films were first prepared, and osteoblast-like and fibroblast cells (bone and tissue cells) were cultured on these films. Based on cell response it was concluded that PVA/halloysite nanotubes film has a high bioactivity with respect to these cells. On top of its biocompatibility, adding halloysite nanotubes into the PVA has also increased the mechanical performance of PVA film. These results indicate that PVA/halloysite nanotubes have a high potential for applications in bone tissue engineering.

An addition of support materials in catalyst's structure leads to an increase in their catalytic performance, and halloysite is a good candidate for support material. Zhang et al. have synthesized palladium (Pd) nanoparticles on silanized halloysite nanotubes and tested the synthesized catalyst in the hydrogenation of styrene (Zhang et al. 2013). The conversion of styrene into ethylbenzene was completed over silanized Pd/halloysite nanotubes in half an hour. In another study, Ru catalysts were supported by natural halloysite nanotubes, and the performance of each catalyst was studied in the decomposition of ammonia (Wang et al. 2011). Highest conversion of ammonia was achieved in the presence of a Ru/halloysite nanotube catalyst with 2.6 wt% Ru loading. Vinokurov et al. have investigated the catalytic performance of cobalt based catalysts in hydrogen production from sodium borohydride (Vinokurov et al. 2018). They have utilized azine or silane as a ligand agent, thereby achieving the adsorption of cobalt (Co) particles onto/into halloysite nanotubes. They have reported that the Co/halloysite nanotube catalyst has a good activity for hydrogen production, and ketazine as a complexing agent increases the adsorption of Co particles onto/into halloysite nanotubes.

CONCLUSION

In conclusion, this book chapter argued the main structural properties and applications of halloysite from previous studies. The occurrence of halloysite was also summarized. Depending on the geological conditions halloysite can occur in various morphologies. The halloysite mineral from Balıkesir, Turkey was structurally characterized, and the results confirm previously published aspects about tubular halloysite. The structural properties of halloysite enable its usage in various fields. Especially, the tubular form and high specific surface area provide loading the halloysite nanotubes with other molecules for pharmaceutical, polymeric and catalytic applications.

REFERENCES

Abdullayev, E., Abbasov, V., Tursunbayeva, A., Portnov, V., Ibrahimov, H., Mukhtarova, G. and Lvov, Y. (2013). Self-healing coatings based on halloysite clay polymer composites for protection of copper alloys. *ACS Applied Materials & Interfaces* 5 (10): 4464–71.

Bailey, S. W. (1990). Halloysite- a critical assessment. *Sciences Geologiques-Memoires* 86 (56): 89–98.

Chen, C. C., Liu, Y. C., Wu, C., Yeh, C., Su, M. and Wu, Y. (2005). Preparation of fluorescent silica nanotubes and their application in gene delivery. *Advanced Materials* 17 (4): 404–7.

Cheng, H., Liu, Q., Yang, J., Zhang, J. and Frost, R. L. (2010). Thermal analysis and infrared emission spectroscopic study of halloysite–potassium acetate intercalation compound. *Thermochimica Acta* 511 (1–2): 124–28.
Churchman, G. J. (2006). Intercalation method using formamide for differentiating halloysite from kaolinite. *Clays and Clay Minerals* 32 (4): 241–48.
Churchman, G. J., Davy, T. J., Aylmore, L. A. G., Gilkes, R. J. and Self, P. G. (1995). "Characteristics of fine pores in some halloysites." *Clay Minerals* 30 (2): 89–98.
De Souza Santos, P., De Souza Santos, H. and Brindley, G. W. (1966). Mineralogical studies of kaolinite-halloysite clays: Part IV. A platy mineral with structural swelling and shrinking characteristics. *American Mineralogist* 51 (11–12): 1640–48.
Ding, F., Larsson, P., Larsson, J. A., Ahuja, R., Duan, H., Rosen, A. and Bolton, K. (2008). The importance of strong carbon-metal adhesion for catalytic nucleation of single-walled carbon nanotubes. *Nano Letters* 8 (2): 463–68.
Drits, V. A., Sakharov, B. A. and Hillier, S. (2018). Phase and structural features of tubular halloysite (7 Å). *Clay Minerals* 53 (4): 691–720.
Ece, Ö. I., Schroeder, P. A., Smilley, M. J. and Wampler, J. M. (2008). Acid-sulphate hydrothermal alteration of andesitic tuffs and genesis of halloysite and alunite deposits in the Biga Peninsula, Turkey. *Clay Minerals* 43 (2): 281–315.
Ece, Ö. I. and Schroeder, P. A. (2007). Clay mineralogy and chemistry of halloysite and alunite deposits in the Turplu area, Balikesir, Turkey. *Clays and Clay Minerals* 55 (1): 18–35.
Fakhrullina, G. I., Akhatova, F. S., Lvov, Y. M. and Fakhrullin, R. F. (2015). Toxicity of halloysite clay nanotubes in vivo: A caenorhabditis elegans study. *Environmental Science: Nano* 2 (1): 54–59.
Falcón, J. M., Sawczen, T. and Aoki, I. V. (2015). Dodecylamine-loaded halloysite nanocontainers for active anticorrosion coatings. *Frontiers in Materials* 2 (69): 1-13.

Garcia Garcia, F. J., Rodríguez, S. G., Kalytta, A. and Reller, A. (2009). "Study of natural halloysite from the Dragon Mine, Utah (USA)." *Zeitschrift Fur Anorganische Und Allgemeine Chemie* 635 (4–5): 790–95.

Hart, R. D., Gilkes, R. J., Siradz, S. and Singh, B. (2002). The nature of soil kaolins from Indonesia and Western Australia. *Clays and Clay Minerals* 50 (2): 198–207.

Ismail, H., Pasbakhsh, P., Fauzi, M. N. A. and Bakar, A. A. (2008). Morphological, thermal and tensile properties of halloysite nanotubes filled ethylene propylene diene monomer (EPDM) nanocomposites. *Polymer Testing* 27 (7): 841–50.

Jagnow, D. H., Hill, C. A. Davis, D. G., Duchene, H. R., Cunningham, K. I., Northup, D. E. and Queen, J. M. (2000). History of the sulfuric acid theory of speleogenesis in the Guadalupe Mountains, New Mexico. *Journal of Cave and Karst Studies* 62 (2): 54–59.

Jia, Z., Luo, Y., Guo, B., Yang, B., Du, M. and Jia, D. (2009). Reinforcing and flame-retardant effects of halloysite nanotubes on LLDPE. *Polymer - Plastics Technology and Engineering* 48 (6): 607–13.

Joussein, E. (2016). Geology and mineralogy of nanosized tubular halloysite. *Developments in Clay Science*. Elsevier B. V., 7: 12-48.

Joussein, E., Petit, S., Churchman, J., Theng, B., Righi, D. and Delvaux B. (2005). Halloysite Clay Minerals — a Review." *Clay Minerals* 40 (4): 383–426.

Levis, S. R. and Deasy, P. B. (1908). Predatory politics in Oklahoma. *Science* 27 (695): 675–76.

Liu, M., Wu, C., Jiao, Y., Xiong, S. and Zhou, C. (2013). Chitosan-halloysite nanotubes nanocomposite scaffolds for tissue engineering. *Journal of Materials Chemistry B* 1 (15): 2078–89.

Lun, H., Ouyang, J. and Yang, H. (2014). Natural halloysite nanotubes modified as an aspirin carrier. *Royal Society of Chemistry Adances* 4 (83): 44197–202.

Margenot, A. J., Calderón, F. J., Goyne, K. W., Mukome, F. N. D. and Parikh, S. J. (2017). IR spectroscopy, soil analysis applications. *Encyclopedia of Spectroscopy and Spectrometry*, 448–54. Elsevier.

Murray, H. H. (1991). Overview-clay mineral applications. *Applied Clay Science* 5: 379-395.

Noro, H. (1986). Hexagonal platy halloysite in an altered tuff bed, Komaki City, Aichi Prefecture, Central Japan. *Clay Minerals* 21 (3): 401–15.

Ouyang, J., Zhou, Z., Zhang, Y. and Yang, H. 2014. High morphological stability and structural transition of halloysite (Hunan, China) in heat treatment. *Applied Clay Science* 101: 16–22.

Parham, W. E. (1969). *Halloysite-rich tropical weathering products of Hong Kong*. Jerusalem: Israel Universities Press.

Plançon, A. (2001). Order-disorder in clay mineral structures. *Clay Minerals* 36 (1): 1–14.

Polyak, V. J. and Güven, N. (2000). Clays in caves of the Guadalupe Mountains, New Mexico." *Journal of Cave and Karst Studies* 62 (2): 120–26.

Ross, C. S. and Kerr, P. F. 1934. Halloysite and allophane. *Professional Paper*, 135–48.

Saklar, S., Ağrılı, H., Zimitoğlu, O., Başara, B. and Kaan, U. (2012). The characterization studies of the Northwest Anatolian halloysites/ kaolinites. *Bulletin Of The Mineral Research and Exploration* 145 (145): 48–61.

Shchukin, D. G., Lamaka, S. V., Yasakau, K. A., Zheludkevich, M. L., Ferreira, M. G. S. and Möhwald, H. (2008). Active anticorrosion coatings with halloysite nanocontainers. *The Journal of Physical Chemistry C* 112 (4): 958–64.

Sudarshan, K., Rath, K. S., Patri, M., Sachdeva, A. and Pujari, P. K. (2007). Positron annihilation spectroscopic studies of fluorinated ethylene propylene copolymer-g-polystyrene. *Polymer* 48 (21): 6434–38.

Tan, D., Yuan, P., Annabi-Bergaya, F., Yu, H., Liu, D., Liu, H. and He, H. (2013). Natural halloysite nanotubes as mesoporous carriers for the loading of ibuprofen. *Microporous and Mesoporous Materials* 179: 89–98.

Thorpe, M. F., Tománek, D. and Enbody, R. J. (2002). Science and application of nanotubes. *Fundamental Materials Research*. Boston, MA: Springer US.

Vergaro, V., Abdullayev, E., Lvov, Y. M., Zeitoun, A., Cingolani R., Rinaldi, R. and Leporatti, S. (2010). Cytocompatibility and uptake of halloysite clay nanotubes. *Biomacromolecules* 11 (3): 820–26.

Vinokurov, V., Stavitskaya, A., Glotov, A., Ostudin, A., Sosna, M., Gushchin, P., Darrat, Y. and Lvov, Y. (2018). Halloysite nanotube-based cobalt mesocatalysts for hydrogen production from sodium borohydride. *Journal of Solid State Chemistry* 268: 182–89.

Wang, L., Chen, J., Ge, L., Zhu, Z. and Rudolph, V. (2011). Halloysite-nanotube-supported Ru nanoparticles for ammonia catalytic decomposition to produce CO_x-free hydrogen. *Energy and Fuels* 25 (8): 3408–16.

White, G. N. and Dixon, J. B. (2002). Kaolin-serpentine minerals: Soil Mineralogy with Environmental Applications. *Soil Science Society of America*, 7: 389–414.

Wilson, I. R., De Souza Santos, H., and De Souza Santos, P. (2006). Kaolin and halloysite deposits of Brazil. *Clay Minerals* 41 (3): 697–716.

Wilson, I. R. (2004). Kaolin and halloysite deposits of China. *Clay Minerals* 39 (1): 1–15.

Wilson, I. and Keeling, J. (2016). Global occurrence, geology and characteristics of tubular halloysite deposits. *Clay Minerals* 51 (3): 309–24.

Yendluri, R., Otto, D. P., De Villiers, M. M., Vinokurov, V. and Lvov, Y. M. (2017). Application of halloysite clay nanotubes as a pharmaceutical excipient. *International Journal of Pharmaceutics* 521 (1–2): 267–73.

Yuan, P., Tan, D. and Annabi-Bergaya, F. (2015). Properties and applications of halloysite nanotubes: Recent research advances and future prospects. *Applied Clay Science* 112–113: 75–93.

Zhang, Y., He, X., Ouyang, J. and Yang, H. (2013). Palladium nanoparticles deposited on silanized halloysite nanotubes: Synthesis, characterization and enhanced catalytic property. *Scientific Reports* 3 (29): 1–6.

Zhou, W. Y., Guo, B., Liu, M., Liao, R., Rabie, A. B. M. and Jia, D. (2010). Poly(vinyl alcohol)/halloysite nanotubes bionanocomposite films: Properties and in vitro osteoblasts and fibroblasts response. *Journal of Biomedical Materials Research - Part A* 93 (4): 1574–87.

In: Halloysite
Editor: Herbert A. Eckart

ISBN: 978-1-53616-812-9
© 2020 Nova Science Publishers, Inc.

Chapter 2

HALLOYSITE-BASED NANOPEAPOD COMPOSITES: RAPID AND CONTROLLED *IN-SITU* GROWTH OF NOBLE METAL NANOSTRUCTURES WITHIN HALLOYSITE CLAY NANOTUBES

*Taha Rostamzadeh, Md Shahidul Islam Khan
and John B. Wiley*[*]

Department of Chemistry and Advanced Materials Research Institute,
University of New Orleans, New Orleans, LA, US

ABSTRACT

Incorporation of nanoparticles (NPs)/nanomaterials into the hollow space of naturally available tubular nanoclays can lead to the creation of advanced functional nanopeapod (NPP) composites. In this chapter, we present details on a simple, rapid (≤ 2 min), high-yield, and low-temperature synthetic approach for the *in-situ* growth of noble metal NPs

---

[*] Corresponding Author's E-mail: jwiley@uno.edu.

in the interior of halloysite nanotubes (HNTs). Proper combination of metal precursors in mixed polar-nonpolar solvents, surfactants and reducing agents with mild heating readily leads to the *in-situ* formation of Au, Ag, and Pd nanostructures within HNTs. The sizes of NPs are tuned mainly by adjusting nucleation and growth rates. Further modification of the process, through an increase in the amount of reducing agent, allows for the formation of nanorods (NRs)/nanowires (NWs) within the HNTs. The ability to readily grow such core–shell nanocomposites advances several applications of HNTs in catalysis, nanoreactors, water treatment, as well as in drug delivery/release systems.

Keywords: halloysite, nanotubes, nanoparticles, *in-situ* growth, noble metals

INTRODUCTION

Naturally occurring halloysite nanotubes (HNTs) and HNTs-based composite structures have been the focus of much scientific literature due to their unique structural features and properties, which offer potential applications in anticorrosive/antibacterial treatments, catalysis, drug delivery systems, and water treatment [1–9]. HNTs are abundant in nature and found in almost pure states; this allows for them to be commercially available and comparatively cheaper than other classes of nanotubes [10]. Depending on their geological location and formation processes, their morphological features, including their lengths, and their inner and outer diameters, are roughly estimated to be between 200–1200, 12–30, and 50–60 nm, respectively [2, 11]. HNTs chemical formula is $Al_2(OH)_4Si_2O_5 \cdot nH_2O$, n = 1–2. Layers of these tubes are usually formed by rolling/scrolling of aluminosilicates sheet [1]. HNTs's inner surface is positively charged due to the presence of alumnilol (Al-OH) group, while their outer surface is charged negatively because of silanol (Si-OH) group. This dual charge character and unique chemistry has made HNTs useful candidates for loading desired materials/entities/components within inner sites or on outer surfaces [12]. Several studies in the area of HNT nanocomposites has thus been dedicated to the formation and/or attachment of noble metal/metal oxide NPs within the tubular structure and/or on the outer surfaces of HNTs [12–15]. Unique features of HNTs combined with complementary properties of NPs may lead to the development of novel nanostructures with superior properties and advanced applications such as those reported in catalysts and drug delivery systems [16, 17].

Superior pH sensitivity and photocatalytic activity for the degradation of organic pollutants were, for example, achieved by the attachment of TiO_2 NPs onto the surface of HNTs via solvothermal treatment [15]. In another work, Wang et al. have reported the formation of Ru/HNT nanocomposites using a wet impregnation method. The obtained Ru/HNT nanocomposite structures were shown to exhibit a superior catalytic activity for the decomposition of ammonia [18]. Vinukurov et al. have also demonstrated the formation of Ru NPs inside HNTs using Schiff bases approach. The capability of the formed nanocomposites for hydrogenation of aromatic compound have also been revealed [19]. Glotov et al. have likewise described the hydrogenation of diverse aromatics by means of HNTs-based composites synthesized through microwave-assisted methods and using various organic ligands [20]. An increase in catalytic performance of salinized HNTs composites formed via Pd NPs deposition on the outer surface of nanotubes were furthermore reported [21]. Hamdi et al., recently reported the insertion of Pd NPs within HNTs and studied their superior catalytic activity for Suzuki-Miyaura cross-coupling reactions, not accessible from individual nanocomposite components [22]. Enhanced catalytic properties and application of Au/ HNTs nanocomposites, synthesized through decomposition/ precipitation process, in the oxidation of benzyl alcohol to benzaldehyde has also been reported by Fu et al. [17]. Additionally, antimicrobial properties of silver (Ag), as another noble metal, grown within HNTs has also been confirmed by Lvov et al. [23].

Functionalized HNTs composites have also been applied to drug delivery systems. Drug delivery in bio-systems, vital for saving lives, is nevertheless challenging for several reasons such as the toxicity of drugs carriers and possible side effects [24]. Bio-compatible HNTs with minor toxicity and no reported side effects, to the best of authors' knowledge, have been displayed as promising drug careers in drug delivery and release systems [25]. Massaro et al., for instance, demonstrated the prospects of HNTs-based nanocomposites as drugs carriers. They have functionalized outer surface of HNTs with cyclodextrin component through microwave irradiation and filled inner lumen with clotrimazole, as an antifungal drug. Kinetic release of this drug onto the targeted cells has then been investigated *in vivo* [26]. The applications of Au NRs @HNTs in drug delivery and release have also been recently investigated by Zhang et al. These functionalized nanostructures loaded with drugs, doxorubicin (DOX) with bovine serum albumin on the outer surface, were shown to be compatible for the treatment of breast cancer cells, while having minimal impact on normal cells. Researchers found that upon chemo-photothermal treatment, the composite exhibited notable tumor-targeted efficiency and good controlled-release effects for DOX [27].

Due to the various important applications of noble metal@HNTs, several synthetic approaches for their formation have been developed. Many of these techniques, however, are time consuming processes and necessitate several steps, special instrumentation, and high temperature [23, 28–30]. In this chapter, we highlight a general strategy for the *in-situ* growth of noble metal NPs and/or NRs/NWs within the HNTs. These techniques can be rapidly applied to the formation of Au@HNTs, Ag@HNTs, and Pd@HNTs nanocomposites. The importance of reaction parameters on the control of NP size and morphological features within the noble metal nanostructures@HNTs will also be discussed.

RESULTS

Rapid *In-Situ* Formation of Gold within HNTs

HNTs can be used as a template for the *in-situ* growth of noble metal NPs [11]. Gold NPs grow directly inside HNTs via a straightforward chemical reduction. In a typical synthesis, HNT templates are combined with gold (III) chloride trihydrate gold precursor, $HAuCl_4 \cdot 3H_2O$, toluene and ethanol as solvents, and oleic acid (OAc) and oleylamine (OAm) surfactants.

These components were mixed and sonicated to obtain a homogeneous mixture. Ascorbic acid (AA) was selected as the main reducing agent, though OAm can also act as a mild reducing agent. The *in-situ* growth of gold NPs inside HNTs was expected due to the capillary force at nanoscale and thus presence of gold ions and other components within the HNTs [1]. The mixture was stirred with a constant temperature (55°C) throughout different processing steps to produce a uniform product. After adding AA, the color of the reaction system changes from orange to dark purple, indicating the formation of gold NPs. The formation of gold NPs within HNTs can be followed by looking at the color change of the solution that can provide an insight into the reduction degrees of gold ions as well as nucleation and growth progresses. Different experimental conditions require dissimilar reaction times for the formation of gold NPs and thus different times for color change. Figure 1, for instance, shows continuous visible change of the reaction system after adding AA (10 mg), to the stirring mixture of HNTs (15mg), $HAuCl_4 \cdot 3H_2O$ (22mg), toluene (2mL), ethanol (2mL), OAm (0.5mL), and OAc (0.5mL).

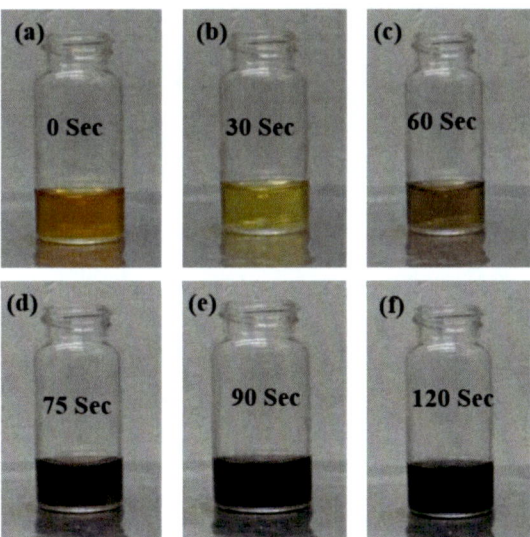

Figure 1. Continuous change of color of the mixture: (a) before adding AA and (b-f) after adding the reducing agent, AA (10 mg).

The color of the system turns from orange to pale yellow within 30 seconds, and then deep purple, after 2 minutes, which appears black due to the high concentration of gold NPs in the colloid. Detailed analysis of the products showed that not only Au NPs can directly grow within HNTs but that they also freely form in solution. TEM images of the products (Figure 2c, d) show Au NPs form in the absence and presence of HNTs and that the product before purification consists of NPs and Au@HNT NPPs.

A purification step is thus required to separate Au@HNT nanocomposite peapods from the free Au NPs. This is accomplished through multiple centrifugation steps with a mixture of ethanol and toluene, segregating the free Au NPs to the supernatant. Figure 3a-e shows TEM images of the separated nanopeapods (NPPs), and gold NPs inside HNTs obtained using 10 mg AA and 4 mL solvents. Detailed examination of the separated nanocomposites show the formation of 6.7 ± 1.7 nm Au NPs with a narrow size distribution (Figure 3f). These images confirm the rapid (2 min.) and high yield formation of Au NPs within HNTs.

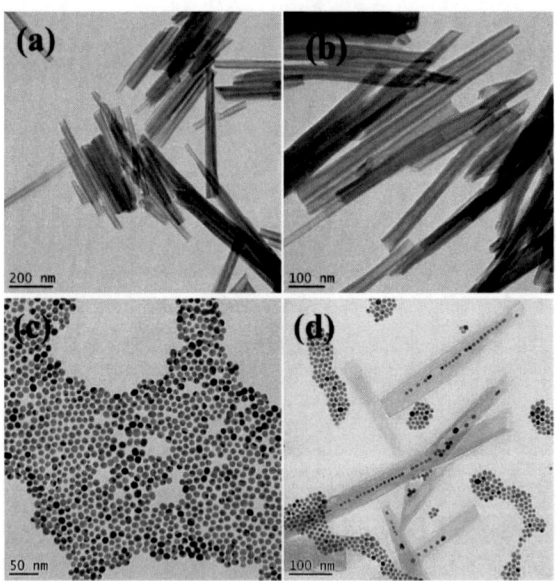

Figure 2. (a-b) TEM images of HNTs. (c) Free Au NPs formation in the absence of HNTs. (d) Au NPs inside and outside of HNTs before purification steps.

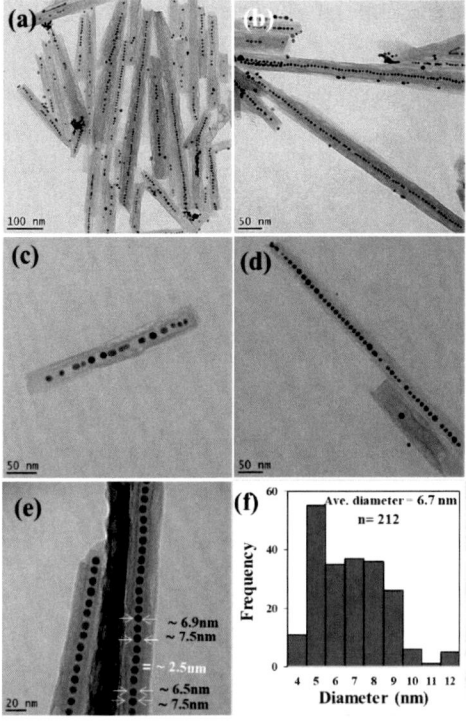

Figure 3. (a-e) presence of Au NPs inside HNTs with diverse morphological features; 10 mg AA and 4 mL of solvent was used for the preparation of NPPs.; (f) histogram of NPs size distribution.

The filling fraction of the separated products, ratio of NPs to HNT available space, is greater more than ~65%. Formation of Au NPs inside HNTs with different sizes and morphological features can also be seen (Figure 3a-d). In a uniformly formed chain of NPs, space between NPs is estimated to be ~ 2.5 ± 0.2 nm (Figure 3e), likely due to the presence of OAm on NP surfaces, similar to that reported by Adireddy et al. [31]. Notably, in some occurrences, double chain of NPs can be observed inside HNT lumen (Figure 3b), with variations in NPs size and inter-particle distances. Size, type, and morphology of the noble metal nanostructures grown within tubular structures can, however, be mainly controlled via nucleation and growth rates and, for instance, by changing the amount/type of reducing agents and solvents.

Impact of Reducing Agents

An increase in the amount of AA, used as the main reducing agent, has a direct impact on the shape and size of Au nanostructures formed inside HNTs (Figure 4 a-d). When the amount of AA is increased from 10 mg to 150 mg, while keeping the other experimental conditions constant, the reaction time was less than 2 minutes and the main product consisted of a combination of gold NRs, NWs, and NPs within HNTs. NPs aggregation are also clearly seen adjacent to the tubular structures (Figure 4 a, b)

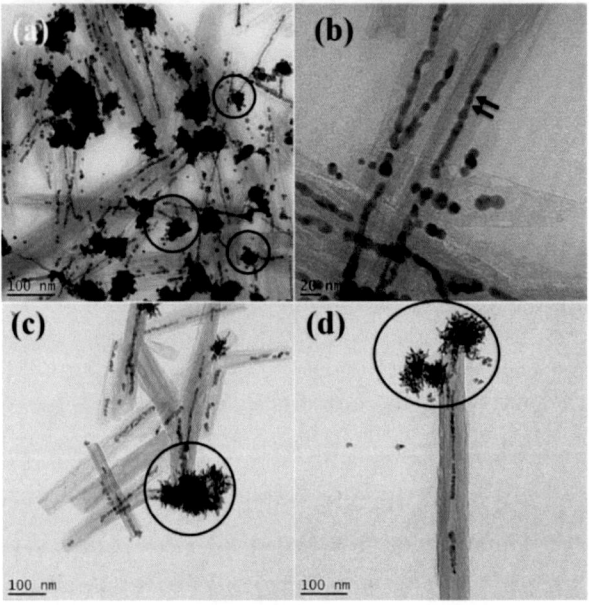

Figure 4. (a, b) Formation of Au nanostructures inside HNTs when (a, b) 150 mg, and (c, d) 75 mg AA and 4 mL of solvent was used. Circles show the formation of aggregates in the entrance of HNTs, and arrows in (b) shows the granular-structure of NRs.

Decreasing the amount of AA to 75 mg resulted in the presence of NPs, aggregates, and NRs, while the ratio of formed NRs to NPs is lower than when 150 mg is used (Figure 4 c, d). It is interesting to note that when HNTs are not included in the same reaction system, clusters of NPs, NRs or NWs were not formed; only free Au NPs seen in the products. This further highlights the role of HNTs as a template for the NP attachment and formation of NRs/NWs.

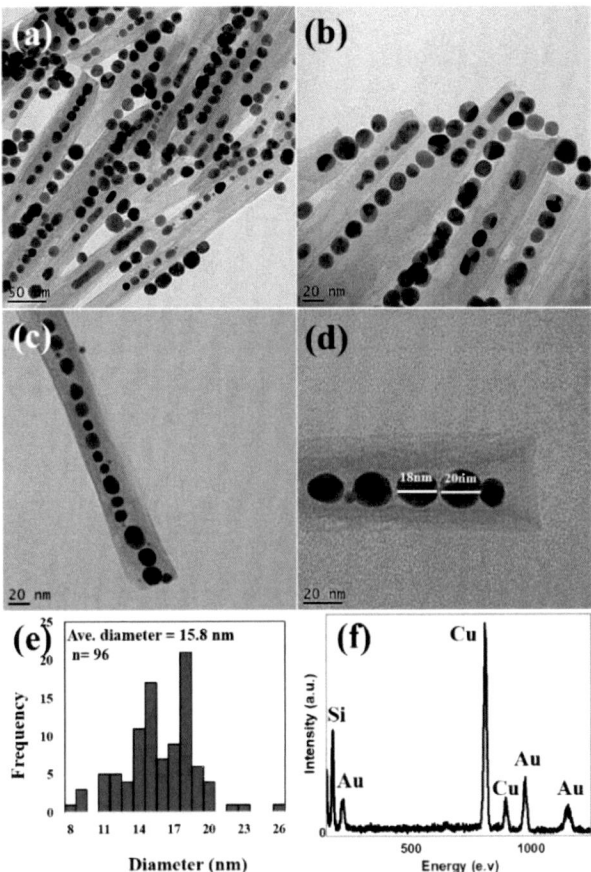

Figure 5. (a-d) TEM images show NPs size clearly increase in the absence of AA and using 4 mL of solvent; (e) histogram shows the size distribution of NPs obtained after 24 h of reaction, and (f) EDS confirms the presence of Au from NPs, Si from HNTs, and Cu from copper grid.

In the absence of AA, with only OAm used as a weak reducing agent, the formation of NPs inside HNTs can also be obtained, though it requires a long reaction, 24 hours. TEM images (Figure 5 a-d) show the formation of

Au NPs with a broad size distribution and an average diameter of ~15.8 nm (Figure 5e) within HNTs. A summary of size and morphological control using different amount of AA is depicted in Figure 6.

Impact of Concentration

In order to explore the influence of concentration on the size and formation of Au NPs, the total amount of the solvents was increased from the typical reaction conditions, 4 mL (2 mL ethanol and 2 mL toluene) to 20 mL (10 mL ethanol and 10 mL toluene), while keeping the amount of surfactants and reducing agent (10 mg) constant. TEM images display that suitable conditions for the formation of larger particles are provided upon a decrease in reactant concentration, via an increase in the amounts of solvents, where NPs size increased from an average of ~ 6.7 nm (Figure 3) to ~ 7.9 nm (Figure 7). Additionally, formation of NRs and aggregated NPs within/adjacent to HNTs entrance can be observed (Figure 7a-c).

In-situ Formation of Silver and Palladium within HNTs

Ag NPs/NWs can also be directly synthesized inside HNTs. This was similarly achieved through a chemical reduction of silver nitrate using AA as the reducing agent. The reaction time for the formation of Ag NPs was higher (~ 0.5 h) than Au NPs (~2 min.) obtained using comparable amount of AA and experimental conditions, likely due to the higher stability of Ag than Au ions presented in the mixtures. Ag NPs grown inside HNTs are more polydispersive and have a broad size distribution, ranging from 2.8 nm to 14 nm. Formation of Ag NWs is also clear (Figure 8 a, b). Notably, the yield of Ag@HNTs is lower than Au@HNTs nanocomposites which can possibly/partly be related to the reduction ability of the synthesis media and adjustable parameters.

Halloysite-Based Nanopeapod Composites 27

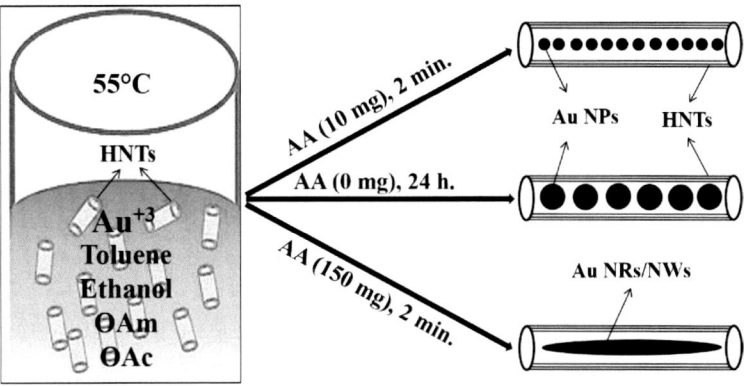

Figure 6. Schematic showing the effects of reducing agent's amount on the size and morphological control of obtained nanocomposites.

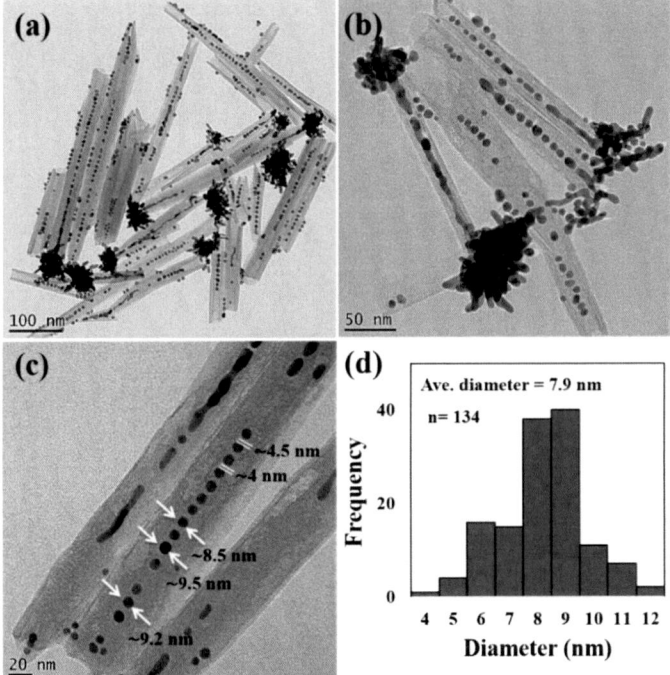

Figure 7. (a-c) TEM images demonstrate the formation of Au NPs/NRs@HNTs after 12 h, when 20 mL of solvents and 10 mg of AA was used. (d) histogram represents the size distribution of obtained NPs with an average size of ~ 7.9 nm.

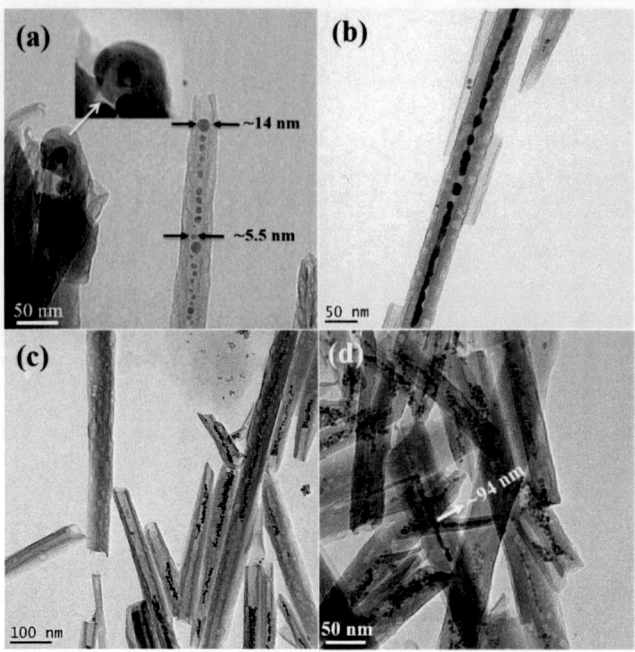

Figure 8. (a-b) TEM images of Ag@HNTs nanocomposites; (a) presence of NPs inside a HTN highlighted by a top-view image in the inset; (b) formation of NWs inside a HNT; (c-d) TEM images of Pd@HNTs nanocomposites consist of mainly Pd NPs (c), and Pd NWs (d) inside HNTs.

Chemical reduction process furthermore allows obtaining Pd nanostructures in HNTs interior. Due to high chemical stability of Pd ions in the system, NaBH$_4$ was used as a strong reducing agent. Compare to Au and Ag, Pd NPs are not very monodisperse. However, spherical NPs have been detected along with NWs. The sizes of spherical Pd NPs is ~ 3 nm on average (Figure 8c). NWs up to ~94 nm in length were also grown inside HNTs (Figure 8d).

DISCUSSION

Noble metal NPs@HNTs nanopeapod composites are expected to unveil new properties not only due to their additive intrinsic properties but also because of the NPs and tubular structures and/or NPs-NPs interactions [32–35]. Rapid formation of noble metal NPs within biocompatible, cheap, and readily available HNTs can facilitate several applications of HNTs-based nanocomposites in catalysis, drug loading, delivery, and release systems, removal and decomposition of organic pollutants, and in water treatments [16, 17, 36, 37]. These tubular structures can be combined with noble metal NPs using many approaches such as surface functionalization and followed by NP attachment, solvent evaporation and NP insertion, solvothermal encapsulation and/or in particular *in-situ* growth of NPs [11, 31, 34, 35, 38].

In-situ growth of noble metal NPs inside HTNs was carried out through a facile, low temperature, and rapid synthetic strategy. OAc and OAm were used as surfactants and in addition to HNTs and noble metal precursor, dissolved/dispersed in a mixture of miscible solvents; ethanol as a polar, and toluene as nonpolar. The selected solvents along with a rapid sonication step and uniform magnetic stirring ensure even dispersions of reactants/components throughout the synthesis/synthesis media (Figure1) from/in which uniform nanostructures can be formed; either a polar or nonpolar solvent cannot provide this condition since gold precursor and HNTs are hydrophilic, while OAc and OAm are hydrophobic [39, 40]. After addition of reducing agent (AA in the formation of Au and Ag, or $NaBH_4$ in the formation of Pd) to the system, solutions color quickly changes indicating the reduction of metal ions and inception of NPs formation through nucleation and growth processes. At the beginning of the process, noble metal atomic concentration in the media increases with the reaction time and nucleation starts when it reaches to a critical point [41–43]. Depending on the stability of the nuclei, some of them start to grow by the addition/absorption of noble metal atoms to their surfaces, while others may disappear. Further attachments/adsorptions of the atoms from the bulk/liquid to the surface of growing crystals lead to the formation of noble metal NPs and subjected to the reaction conditions and progresses, atomic concentration of noble metals within the synthesis media may decrease due to their adsorptions to the growing crystals [42–44]. Presented surfactants also interact with the surface of growing crystals and influence their final shape, morphology, and size. Surfactant attachments to the surface of growing crystals/NPs also allow uniform growth of NPs and prevent nanocrystal aggregations [42, 44, 45]. Depending on the reaction conditions, nucleation and growth rates, as well as the thermodynamics of surfactant interactions with and kinetics of their attachments to the surface of growing crystals, different shape, morphology and size of NPs can be obtained [46].

Our experimental designs and characterizations indicate that NPs form both inside the liquid, as homogeneous sites, and directly within the HNTs, as heterogeneous sites (Figure 2). Further, NPs were not inserted within the HNTs using a simplified stirring of HNTs and preformed Au NPs mixed within the solvents. Presence of different NPs size observed for NPs inside and outside of HNTs (Figure 2d) may also be directly related to their homogeneous and heterogeneous formations and the nucleation/growth rates of the crystals [41]. In order to directly grow noble metal NPs inside HNTs, a steady flow of solution containing metal ions/atoms, surfactants, and reducing agents should be presented within the interior of HNTs. Our calculation/estimation for the formation of Au NPs in a NPP, for instance where the total volume of NPs is approximately 15% of the total volume of HNTs interior, also showed that a continuous flow of solution, thousand times more than total interior volume of HNTs, should rapidly enter into HNTs. Introduction of the liquid inside HTNs occurs mainly due to capillary force and can be enhanced when the inner diameter of a tube becomes smaller [30]. Our observations also confirm the presence of more Au NPs in HNTs with smaller inner diameters (Figure 3a, b, and d). Capillary force and thus the introduction rate of solution inside HNTs can be influenced by several factors such as temperature, surface tension, and viscosity [28, 30, 47]. By tuning these parameters not only nanocrystals formation mechanism within HNTs can be investigated, but one may also consider to further study capillary force at nanoscale via facile construction of varieties of noble metal nanostructures directly/easily formable inside tubular structures.

Size and morphological features of noble metal nanostructures (Au, Ag, and Pd) were achieved/are achievable by careful attentions to the reaction parameters (Figure 3-8) and selecting an appropriate reducing agent for the reduction of noble metal ions; AA for Au and Ag and $NaBH_4$ for Pd. High yield formation of ~ 6.7 nm Au NPs were formed after 2 minutes of the reaction where 10 mg AA and 4 mL of solvent used (Figure 3). Beside the formation of NPs, noble metal NRs and NWs were also directly synthesized inside HNTs (Figure 4, 5, 7, and 8). In terms of Au, our designed experiments and characterizations suggest that higher amount of AA used as reducing agent, encourage the formation of NWs/NRs (Figure 4). Very high rate of reduction and thus high rate of nucleation and growth, provided by using a large amount of reducing agent, may decrease the frequency of surfactant attachments to the surface of crystals, throughout/after the nanocrystals formation [48]. This resulted in the formation of NRs/NWs within HNTs as well as aggregated NPs formed near HNT entrances (Figure 4). Detailed examination of Figure 4 also shows that NRs/NWs are formed via a direct growth and/or through coalescence and oriented attachments where adjacent NPs fused together to minimize their interfacial energy through the attachments of random or common crystallographic facets, respectively, and formed granular-shaped NRs or NWs. Formation of aggregates adjacent/attached to the HNT entrances may further display the continuous stream of liquid into the HNTs, due to the capillary force, throughout the nanostructures formation. Interestingly, formation of Au NRs/NWs and aggregates were not detected where HNTs was removed from the same synthesis media. This may indicate the role of HNT as a template for the direct formation of NRs/NWs or via NP attachments. Additionally, the rate of nucleation and growth as well as surfactant interactions/attachments to the surface of growing crystals may differ for the particles formed inside the liquid and those formed directly within HNTs (Figure 6) resulting in the formation of nanostructures with different size as well as morphology.

Noble metal NPs size are controllable by the nucleation and growth rates using different amount of reducing agent and/or solvents. For instance in the synthesis of Au NPPs, the removal of AA from the system and hence decreasing the reduction ability of the synthesis media, using only OAm as a weak reducing agent, results in the formation of larger Au NPs (~15.8 nm) obtained after 24 h (Figure 5). Reduction ability can also be decreased by increasing the amount of solvent [49]. While 10 mg AA presented in the system, an increase in the amount of solvent from 4 mL to 20 mL, leads to an increase in the reaction time from 2 minutes to 12 h and NPs size from ~ 6.7 nm to ~7.9 nm, respectively (Figure 7). Formation of NRs within and aggregates adjacent to HNTs entrance are most likely due to the presence of less surfactants to the solvents in this particular system (Figure 7). A decrease in the reduction ability of these two differentiated synthesis media, outcomes in lower generation rates of gold atoms and thus lower nucleation rates so that fewer nuclei are formed. Since the amount of gold precursor is kept constant, in both systems, larger NPs are obtained after the growth completion. Clear increase in the reaction times can also be directly related to the generation rates of gold atoms and their migration/adsorption rates to the surface of growing crystals [50].

Our developed synthetic approach also allowed the formation of Ag or Pd NPs/NRs/NWs inside HNTs (Figure 8) and can further be established for the size and morphological control of diverse noble metals nanostructures. Several types of precursors, reducing agents, surfactants, solvents, and their combined ratios along with temperature are accessible to control the *in-situ* formation of NPs within HNTs through the heterogeneous nucleation and subsequent growth processes. Diverse reducing agents with different reduction abilities can be preferred to regulate the generation rate of noble metal atoms and thus nucleation and growth rates [51]. Sodium citrate, sodium borohydride, sodium carbonate, sodium hydroxide, oleylamine, hydroxylamine, hydrogen peroxide, hydrochloride, citric acid, carbon monoxide, and hydrogen have been reported to act as reducing agents for the reduction of different noble metal ions to atoms [43, 44, 46]. Generally, a strong reducing agent can increase the nucleation rate and lead to the formation of smaller NPs [48]. Low concentration of solutes as well as the existence of surfactants in different synthesis media can hinder the rapid nucleation and following diffusions of growth species from the surrounding solution/media to the growth surfaces and impede their rapid adsorbtions to the surface of growing crystals, resulting in controlled *in-situ* formation of NPs in tubular structures. Subjected to the surface chemistry, interaction of a surfactant/capping agent with a formed/growing solid/nucleus is one of the main synthetic parameters that can influence NPs size, shape, and surface functionality [52]. For instance, a strong interaction/adsorption of a specific surfactant to the surface of growth sites could reduce the growth rate. Additionally; different surfactants may have dissimilar tendencies towards diverse crystals facets, so that they can encourage/hinder the growth rate of a particular surface, resulting in shape control formations of noble metal NPs. NPs shape, size and surface functionality play a significant role in the properties of the accessible nanopeapod composites including dispersibility, catalytic activity, stability, and reactivity [42–44]. Our rapid and versatile synthetic strategy along with aforementioned and adjustable factors can thus be used to further control structural features of diverse classes of noble metal NPs directly grown inside HNTs and improve their varieties of applications in catalysis, water treatments, and in drug loading, delivery and release systems.

REFERENCES

[1] Lvov, Y; Wang, W; Zhang, L; Fakhrullin, R. Halloysite Clay Nanotubes for Loading and Sustained Release of Functional Compounds. *Adv. Mater.*, 2016, 28 (6), 1227–1250. https://doi.org/10.1002/adma.201502341.

[2] Lvov, YM; Shchukin, DG; Möhwald, H; Price, RR. Halloysite Clay Nanotubes for Controlled Release of Protective Agents. *ACS Nano*, 2008, 2 (5), 814–820. https://doi.org/10.1021/nn800259q.

[3] Du, M; Guo, B; Jia, D. Newly Emerging Applications of Halloysite Nanotubes: A Review; *Poly Int*, 2010, 59, 574-582 https://doi.org/10.1002/pi.2754.

[4] Joussein, E; Petit, S; Churchman, J; Theng, B; Righi, D; Delvaux, B. Halloysite Clay Minerals – a Review. *Clay Miner.*, 2016, 40 (4), 383–426. https://doi.org/10.1180/0009855054040180.

[5] Abdullayev, E; Lvov, Y. Clay Nanotubes for Corrosion Inhibitor Encapsulation: Release Control with End Stoppers. *J. Mater. Chem.*, 2010, 20 (32), 6681–6687. https://doi.org/10.1039/C0JM00810A.

[6] Song, X; Zhou, L; Zhang, Y; Chen, P; Yang, Z. A Novel Cactus-like Fe_3O_4/Halloysite Nanocomposite for Arsenite and Arsenate Removal

from Water. *J. Clean. Prod.*, 2019, 224, 573–582. https://doi.org/10.1016/j.jclepro.2019.03.230.

[7] Xing, X; Xu, X; Wang, J; Hu, W. Preparation, Release and Anticorrosion Behavior of a Multi-Corrosion Inhibitors-Halloysite Nanocomposite. *Chem. Phys. Lett.*, 2019, 718, 69–73. https://doi.org/10.1016/j.cplett.2019.01.033.

[8] Song, S; Zhao, T; Qiu, F; Zhu, W; Wu, Y; Ju, Y; Dong, L. Silver Nanoparticle Decorated Halloysite Nanotube for Efficient Antibacterial Application. *Chem. Phys.*, 2019, 521, 51–54. https://doi.org/10.1016/j.chemphys.2019.01.020.

[9] Stavitskaya, A; Batasheva, S; Vinokurov, V; Fakhrullina, G; Sangarov, V; Lvov, Y; Fakhrullin, R. Antimicrobial Applications of Clay Nanotube-Based Composites. *Nanomaterials.*, 2019, 9 (5), 708. https://doi.org/10.3390/nano9050708.

[10] Nazir, MS; Kassim, MHM; Mohapatra, L; Gilani, MA; Raza, MR; Majeed, K. Characteristic Properties of Nanoclays and Characterization of Nanoparticulates and Nanocomposites. In *Nanoclay Reinforced Polymer Composites*, Springer, 2016, pp 35–55.

[11] Rostamzadeh, T; Islam Khan, MS; Riche', K; Lvov, YM; Stavitskaya, AV; Wiley, JB. Rapid and Controlled *In Situ* Growth of Noble Metal Nanostructures within Halloysite Clay Nanotubes. *Langmuir*, 2017, 33 (45), 13051–13059. https://doi.org/10.1021/ acs.langmuir.7b02402.

[12] Zhu, H; Du, M; Zou, M; Xu, C; Fu, Y. Green Synthesis of Au Nanoparticles Immobilized on Halloysite Nanotubes for Surface-Enhanced Raman Scattering Substrates. *Dalton Trans.*, 2012, 41 (34), 10465–10471. https://doi.org/10.1039/C2DT30998J.

[13] Fu, Y; Zhang, L. Simultaneous Deposition of Ni Nanoparticles and Wires on a Tubular Halloysite Template: A Novel Metallized Ceramic Microstructure. *J. Solid State Chem.*, 2005, 178 (11), 3595–3600. https://doi.org/10.1016/j.jssc.2005.09.022.

[14] Vinokurov, VA; Stavitskaya, AV; Chudakov, YA; Ivanov, EV; Shrestha, LK; Ariga, K; Darrat, YA; Lvov, YM. Formation of Metal Clusters in Halloysite Clay Nanotubes. *Sci. Technol. Adv. Mater.*

2017, 18 (1), 147–151. https://doi.org/10.1080/14686996.2016.1278352.

[15] Wang, R; Jiang, G; Ding, Y; Wang, Y; Sun, X; Wang, X; Chen, W. Photocatalytic Activity of Heterostructures Based on TiO_2 and Halloysite Nanotubes. *ACS Appl. Mater. Interfaces*, 2011, 3 (10), 4154–4158. https://doi.org/10.1021/am201020q.

[16] Dzamukova, MR; Naumenko, EA; Lvov, YM; Fakhrullin, RF. Enzyme-Activated Intracellular Drug Delivery with Tubule Clay Nanoformulation. *Sci. Rep.*, 2015, 5, 10560. https://doi.org/10.1038/srep10560.

[17] Fu, X; Ding, Z; Zhang, X; Weng, W; Xu, Y; Liao, J; Xie, Z. Preparation of Halloysite Nanotube-Supported Gold Nanocomposite for Solvent-Free Oxidation of Benzyl Alcohol. *Nanoscale Res. Lett.*, 2014, 9 (1), 282. https://doi.org/10.1186/1556-276X-9-282.

[18] Wang, L; Chen, J; Ge, L; Zhu, Z; Rudolph, V. Halloysite-Nanotube-Supported Ru Nanoparticles for Ammonia Catalytic Decomposition to Produce COx-Free Hydrogen. *Energy Fuels*, 2011, 25 (8), 3408–3416. https://doi.org/10.1021/ef200719v.

[19] Vinokurov, VA; Stavitskaya, AV; Chudakov, YA; Glotov, AP; Ivanov, EV; Gushchin, PA; Lvov, YM; Maximov, AL; Muradov, AV; Karakhanov, EA. Core-Shell Nanoarchitecture: Schiff-Base Assisted Synthesis of Ruthenium in Clay Nanotubes. *Pure Appl. Chem.*, 2018, 90 (5), 825–832. https://doi.org/10.1515/pac-2017-0913.

[20] Glotov, A; Stavitskaya, A; Chudakov, Y; Ivanov, E; Huang, W; Vinokurov, V; Zolotukhina, A; Maximov, A; Karakhanov, E; Lvov, Y. Mesoporous Metal Catalysts Templated on Clay Nanotubes. *Bull. Chem. Soc. Jpn.*, 2018, 92 (1), 61–69. https://doi.org/10.1246/bcsj.20180207.

[21] Zhang, Y; He, X; Ouyang, J; Yang, H. Palladium Nanoparticles Deposited on Silanized Halloysite Nanotubes: Synthesis, Characterization and Enhanced Catalytic Property. *Sci. Rep.*, 2013, 3, 2948. https://doi.org/10.1038/srep02948.

[22] Hamdi, J; Blanco, AA; Diehl, B; Wiley, JB; Trudell, ML. Room-Temperature Aqueous Suzuki–Miyaura Cross-Coupling Reactions

Catalyzed via a Recyclable Palladium@Halloysite Nanocomposite. *Org. Lett.*, 2019. https://doi.org/10.1021/acs.orglett.9b00042.

[23] Abdullayev, E; Sakakibara, K; Okamoto, K; Wei, W; Ariga, K; Lvov, Y. Natural Tubule Clay Template Synthesis of Silver Nanorods for Antibacterial Composite Coating. *ACS Appl. Mater. Interfaces.*, 2011, 3 (10), 4040–4046. https://doi.org/10.1021/ am200896d.

[24] Borm, PJ; Müller-Schulte, D. Nanoparticles in Drug Delivery and Environmental Exposure: Same Size, Same Risks? *Nanomed.*, 2006, 1 (2), 235–249. https://doi.org/10.2217/17435889.1.2.235.

[25] Long, Z; Wu, YP; Gao, HY; Zhang, J; Ou, X; He, RR; Liu, M. *In Vitro* and *in Vivo* Toxicity Evaluation of Halloysite Nanotubes. *J. Mater. Chem. B.*, 2018, 6 (44), 7204–7216. https://doi.org/ 10.1039/ C8TB01382A.

[26] Massaro, M; Campofelice, A; Colletti, CG; Lazzara, G; Noto, R; Riela, S. Functionalized Halloysite Nanotubes: Efficient Carrier Systems for Antifungine Drugs. *Appl. Clay Sci.*, 2018, 160, 186–192. https://doi.org/10.1016/j.clay.2018.01.005.

[27] Zhang, J; Luo, X; Wu, YP; Wu, F; Li, YF; He, RR; Liu, M. Rod in Tube: A Novel Nanoplatform for Highly Effective Chemo-Photothermal Combination Therapy toward Breast Cancer. *ACS Appl. Mater. Interfaces.*, 2019, 11 (4), 3690–3703. https://doi.org/ 10.1021/acsami.8b17533.

[28] Ajayan, PM; Lijima, S. Capillarity-Induced Filling of Carbon Nanotubes. *Nature*, 1993, 361 (6410), 333–334. https://doi.org/ 10.1038/361333a0.

[29] FU, Q; Gisela, W; SU, D. Selective Filling of Carbon Nanotubes with Metals by Selective Washing. *New Carbon Mater.*, 2008, 23 (1), 17–20. https://doi.org/10.1016/S1872-5805(08)60008-6.

[30] Ugarte, D; Châtelain, A; Heer, WA. de. Nanocapillarity and Chemistry in Carbon Nanotubes. *Science*, 1996, 274 (5294), 1897–1899. https://doi.org/10.1126/science.274.5294.1897.

[31] Adireddy, S; Carbo, CE; Rostamzadeh, T; Vargas, JM; Spinu, L; Wiley, JB. Peapod-Type Nanocomposites through the *In Situ* Growth of Gold Nanoparticles within Preformed Hexaniobate Nanoscrolls.

Angew. Chem. Int. Ed., 2014, 53 (18), 4614–4617. https://doi.org/10.1002/anie.201310834.

[32] Adireddy, S; Carbo, CE; Yao, Y; Vargas, JM; Spinu, L; Wiley, JB. High-Yield Solvothermal Synthesis of Magnetic Peapod Nanocomposites via the Capture of Preformed Nanoparticles in Scrolled Nanosheets. *Chem. Mater.*, 2013, 25 (19), 3902–3909. https://doi.org/10.1021/cm402352k.

[33] Rostamzadeh, T; Adireddy, S; Wiley, JB. Formation of Scrolled Silver Vanadate Nanopeapods by Both Capture and Insertion Strategies. *Chem. Mater.*, 2015, 27 (10), 3694-3699. https://doi.org/10.1021/acs.chemmater.5b01161.

[34] Adireddy, S; Rostamzadeh, T; Carbo, CE; Wiley, JB. Particle Placement and Sheet Topological Control in the Fabrication of Ag–Hexaniobate Nanocomposites. *Langmuir*, 2014, 31 (1), 480–485. https://doi.org/10.1021/la503775f.

[35] Yao, Y; Chaubey, GS; Wiley, JB. Fabrication of Nanopeapods: Scrolling of Niobate Nanosheets for Magnetic Nanoparticle Chain Encapsulation. *J. Am. Chem. Soc.*, 2012, 134 (5), 2450–2452. https://doi.org/10.1021/ja206237v.

[36] Papoulis, D; Komarneni, S; Panagiotaras, D; Stathatos, E; Toli, D; Christoforidis, KC; Fernández-García, M; Li, H; Yin, S; Sato, T; et al. Halloysite–TiO_2 Nanocomposites: Synthesis, Characterization and Photocatalytic Activity. *Appl. Catal. B Environ.*, 2013, 132–133, 416–422. https://doi.org/10.1016/j.apcatb.2012.12.012.

[37] Yu, L; Wang, H; Zhang, Y; Zhang, B; Liu, J. Recent Advances in Halloysite Nanotube Derived Composites for Water Treatment. *Environ. Sci. Nano*, 2016, 3 (1), 28–44. https://doi.org/10.1039/C5EN00149H.

[38] Rostamzadeh, T; Adireddy, S; Wiley, JB. Formation of Scrolled Silver Vanadate Nanopeapods by Both Capture and Insertion Strategies. *Chem. Mater.*, 2015, 27 (10), 3694–3699. https://doi.org/10.1021/acs.chemmater.5b01161.

[39] Mourdikoudis, S; Liz-Marzán, LM. Oleylamine in Nanoparticle Synthesis. *Chem. Mater.*, 2013, 25 (9), 1465–1476. https://doi.org/10.1021/cm4000476.

[40] Diaz, G; Melis, M; Batetta, B; Angius, F; Falchi, AM. Hydrophobic Characterization of Intracellular Lipids *in Situ* by Nile Red Red/Yellow Emission Ratio. *Micron*, 2008, 39 (7), 819–824. https://doi.org/10.1016/j.micron.2008.01.001.

[41] Porter, DA; Easterling, KE. *Phase Transformations in Metals and Alloys, Third Edition (Revised Reprint)*, CRC Press, 1992.

[42] Cushing, BL; Kolesnichenko, VL; O'Connor, CJ. Recent Advances in the Liquid-Phase Syntheses of Inorganic Nanoparticles. *Chem. Rev.*, 2004, 104 (9), 3893–3946. https://doi.org/10.1021/cr030027b.

[43] Cao, G. *Nanostructures and Nanomaterials: Synthesis, Properties and Applications*, World Scientific, 2004.

[44] Sun, Y; Xia, Y. Shape-Controlled Synthesis of Gold and Silver Nanoparticles. *Science*, 2002, *298* (5601), 2176–2179. https://doi.org/10.1126/science.1077229.

[45] Abécassis, B; Testard, F; Spalla, O; Barboux, P. Probing *in Situ* the Nucleation and Growth of Gold Nanoparticles by Small-Angle X-Ray Scattering. *Nano Lett.*, 2007, 7 (6), 1723–1727. https://doi.org/10.1021/nl0707149.

[46] *Engineering Nanoarchitectures from Nanosheets, Nanoscrolls, and Nanopa* by Taha Rostamzadeh https://scholarworks.uno.edu/td/2229/.

[47] Supple, S; Quirke, N. Rapid Imbibition of Fluids in Carbon Nanotubes. *Phys. Rev. Lett.*, 2003, 90 (21), 214501. https://doi.org/10.1103/PhysRevLett.90.214501.

[48] Thanh, NT; Maclean, N; Mahiddine, S. Mechanisms of Nucleation and Growth of Nanoparticles in Solution. *Chem. Rev.*, 2014, 114 (15), 7610–7630. https://doi.org/10.1021/cr400544s.

[49] Shevchenko, EV; Talapin, DV; Schnablegger, H; Kornowski, A; Festin, Ö; Svedlindh, P; Haase, M; Weller, H. Study of Nucleation and Growth in the Organometallic Synthesis of Magnetic Alloy Nanocrystals: The Role of Nucleation Rate in Size Control of $CoPt_3$

Nanocrystals. *J. Am. Chem. Soc.*, 2003, 125 (30), 9090–9101. https://doi.org/10.1021/ja0299371.

[50] Liu, X; Atwater, M; Wang, J; Dai, Q; Zou, J; Brennan, JP; Huo, Q. A Study on Gold Nanoparticle Synthesis Using Oleylamine as Both Reducing Agent and Protecting Ligand. *J. Nanosci. Nanotechnol.*, 2007, 7 (9), 3126–3133. https://doi.org/10.1166/jnn.2007.805.

[51] Altavilla, Claudia; Enrico, Ciliberto. "Inorganic nanoparticles: synthesis, applications, and perspectives, *CRC Press*, New York. 2011.

[52] Yin, Y; Alivisatos, AP. Colloidal Nanocrystal Synthesis and the Organic-Inorganic Interface. *Nature*, 2005, 437 (7059), 664–670. https://doi.org/10.1038/nature04165.

In: Halloysite
Editor: Herbert A. Eckart

ISBN: 978-1-53616-812-9
© 2020 Nova Science Publishers, Inc.

Chapter 3

THE HALLOYSITE NANOSTRUCTURE-BASED MATERIALS

Ayesha Kausar[*]
National University of Sciences and Technology,
Islamabad, Pakistan

ABSTRACT

The halloysite nanotubes are naturally occurring materials. The halloysite nanotubes are promising candidates for manufacturing polymer-based nanocomposites. The halloysite nanotubes have large surface area and range of exciting nanofiller properties such as mechanical strength, thermal stability, crystallization, and noncytotoxicity.

In this chapter, the processing and properties of various polymer/halloysite nanotube materials have been discussed. Benefits of the halloysite nanotubes for the polymeric structures and technical applications in various fields are also highlighted. Especially, the thermal stability, mechanical strength, and non-flammability of the polymeric nanocomposite have been enhanced using the halloysite addition. The future research is needed to further develop the potential of polymer/ halloysite nanotube materials.

---

[*] Corresponding Author's E-mail: asheesgreat@yahoo.com.

Keywords: Halloysite, polymer, nanocomposite, application

INTRODUCTION

The halloysite is a natural nanoclay i.e., aluminosilicate [1, 2]. It has a hollow nanotubular structure. It consists of SiO_4 tetrahedra and AlO_6 octahedra [3].

Owing to the nano-size, high surface area and length to diameter ratio, halloysite nanotube (HNT) has been used as nanofiller in the polymer composites.

The solution technique, melt route, *in situ* method, and other techniques have been used to incorporate HNT in the polymers [4, 5]. The polymer matrices such as rubbers, poly(methyl methacrylate), polystyrene, polyamide, polyurethane, etc. have been used for the HNT dispersion [6]. The halloysite nanotube and polymer-based nanostructures have significantly enhanced mechanical, thermal, non-flammability, conductivity, anti-corrosion, shape memory, biomedical, and other essential physical properties [7, 8].

The interfacial interactions between the polymer matrices and HNT may enhance the nanofiller dispersion even at high loadings. The research has revealed several applications of polymer/halloysite nanostructures including anti-corrosion coatings, non-flammability materials, shape memory materials, membranes, drug delivery vehicles, and tissue engineering scaffolds [9-12].

In this chapter, the polymer/halloysite nanostructures have been focused in terms of the structure, properties, preparation, applications, and fucture challenges. This comprehensive article also emphases various types of the polymer/halloysite nanostructures.

The purpose of this effort is to reveal the potential of halloysite nanotube with the polymer matrices.

THE HALLOYSITE NANOTUBE: STRUCTURE AND PROPERTIES

The word *halloysite* is resultant of its discoverer *Omaliusd Halloy* (Belgium 1882) [13]. The halloysite is naturally occurring mineral i.e., aluminosilicate clay ($Al_2Si_2O_5(OH)_4nH_2O$) [14]. Its structure is very similar to the kaolinite nanoclay. The halloysite nanotube consists of Al, Si, O and H element. The halloysite occurs in different forms such as nanotube, nanoparticle, nanofiber, etc. [15]. The halloysite nanotube is extratcted from the natural weathered rocks and soils [16]. The HNT may form a single layer or multilayer tubes of varying diamater. Similar to carbon nanotube, the HNT may form single-walled halloylsite nanotube and multi-walled halloylsite nanotube. Single layered nanoclay-based HNT had internal diameter of 10 - 100nm. The specific surface area of halloysite is around 6 m^2/g. It has pore volume of ~1.25mL/g. Its external diameter may vary from 30 - 190nm [5]. Consequently, the HNT possess high aspect ratio, high porosity, non swelling, and regeneration ability. The halloysite nanotube structure possesses outer surface of SiO_2 and inner cylinder core of Al_2O_3. As extracted from natural ores, the HNT is a low cost material which can be used in the mass scale industrial applications. For the drug delivery applications, the halloysite lumen is filled with targeted molecules. The halloysite may retain and release hydrophilic and hydrophobic agents. The HNT has been widely employed as the polymeric nanofiller in the composite materials [17].

THE POLYMER/HALLOYSITE NANOTUBE NANOMATERIALS: DESIGN AND FEATURES

Polyamide has been used used as a matrix material for HNT. The polyamide 11 and halloysite nanotube (PA 11/HNT) nanocomposites have been prepared by the melt extrusion method [18]. The PA 11/HNT has shown excellent transparency, rheology, thermal stability, and mechanical properties. Fine halloysite dispersion in the PA 11 matrix may enhance the ductility, tensile strength, and Young modulus of the nanocomposite. The HNT addition improved the glass transition temperature, crystallization temperature, and degree of crystallinity of the nanocomposites [19]. Lecouvet et. al. [20] formed polyamide 12 (PA 12) and halloysite nanotube nanocomposite using twin-screw mini-compounder. Figure 1 shows transmission electron microscope (TEM) micrograph of halloysite nanotubes. The outer nanotube diameter was in the range of 30-100 nm, while inner lumen diameter was around 15 nm. The aspect ratio was about 10-40. Table 1 shows 5% and 10% weight loss temperatures for neat PA12 and the filled PA12/HNT nanocomposite. The 10 wt.% loaded nanocompsoite showed 31°C higher decomposition temperature than the neat PA12 [21]. The polyamide/halloysite nanotube nanocomposites are promising candidates for the structural applications.

Polystyrene is a transparent and brittle thermoplastic. The polystyrene nanocomposites with different HNT content have been prepared using *in situ* and bulk polymerization of styrene [22].

The thermal stability of PS/HNT nanocomposites was found better than the pure polystyrene.

Figure 1. TEM micrograph of HNT [20].

Table 1. TGA data of internal mixed PA12/HNT nanocomposite in air at 10°C/min. [20]

HNT Content (wt.%)	Temperature at 5% weight loss (°C)	Temperature at 10% weight loss (°C)	Temperature at maximum rate of weight loss (°C)
0	391	403	460
2	404	421	466
5	406	425	468
10	422	434	471

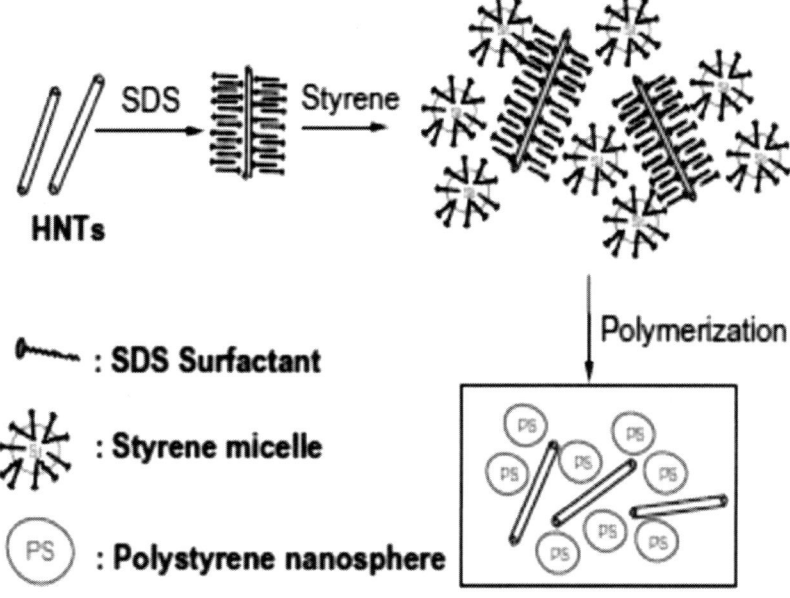

Figure 2. Schematic of *in situ* polymerization of styrene in presence of HNT [24].

The core-shell nanocomposite microspheres of the polystyrene core and halloysite nanotube shell have been developed *via* Pickering suspension polymerization [23]. Lin et. al. [24] designed high impact polystyrene nanocomposite filled with halloysite nanotube using emulsion polymerization. The styrene polymerization was carried out in the presence of sodium dodecyl sulfate (SDS). The SDS acted as a dispersing agent for HNT. Figure 2 shows *in situ* polymerization of styrene in the presence of

HNT. The SDS surfactant developed bilayer on the HNT surface. Thus, SDS formed a uniform and stable suspension. Figure 3a and b reveals the scanning electron microscope (SEM) and transmission electron microscope (TEM) images of HNT. The micrographs showed short tubes and long tubes having length of few to several nm. The aspect ratio of HNT was ranging from 1 to 100. The internal and external diameters of nanotubes were 20 and 50nm, respectively. Figure 4 shows the fracture surface near the edge of an impact sample. The sample depicted the plastic deformation with the dispersed nanotubes.

Chitosan is a linear polysaccharide with β-(1-4)-linked D-glucosamine and N-acetyl-D-glucosamine. It is a biodegradable and biocompatible polymer [25].

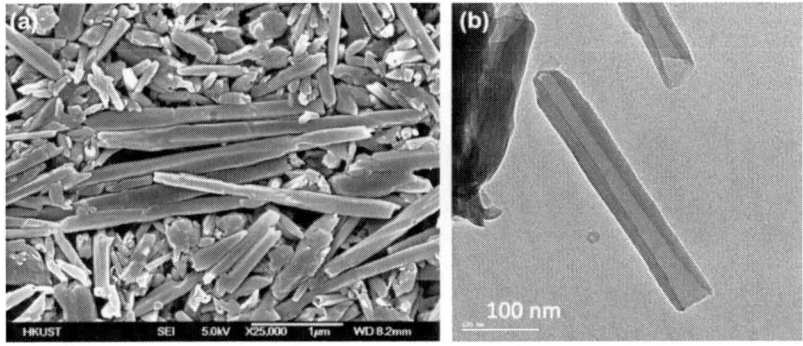

Figure 3. Micromorphology of HNT: (a) SEM and (b) TEM micrographs [24].

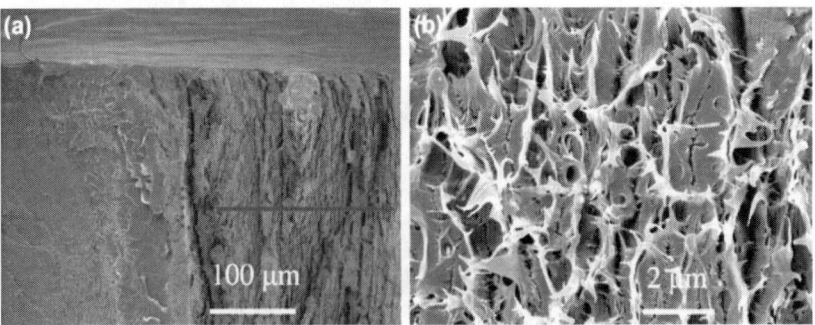

Figure 4. SEM micrographs obtained near edge of fracture surface of PS/HNT nanocomposite [24].

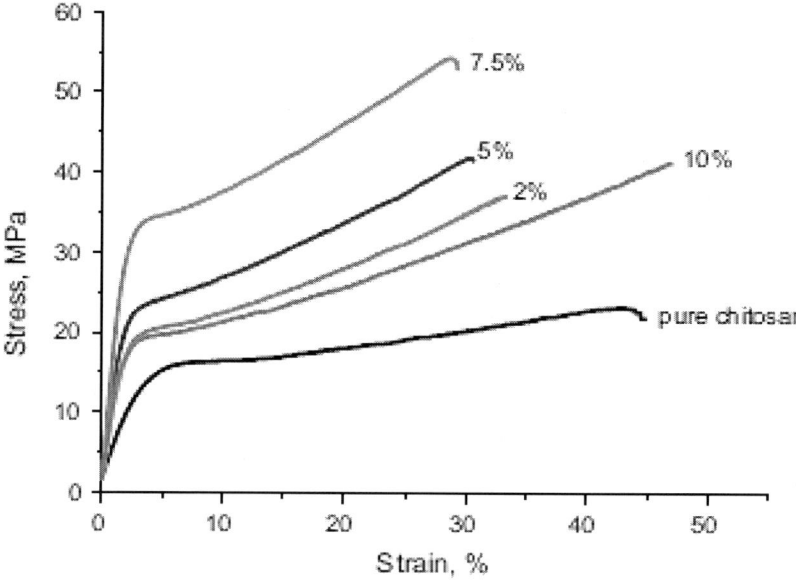

Figure 5. Tensile property for chitosan/HNT nanocomposite films [28].

Chitosan may exist in the form of film, fiber, and hydrogel [26]. The chitosan films have been used to form artificial skin, tissue engineering, and drug carrier [27]. Liu et. al. [28] developed chitosan and halloysite nanotube-based nanocomposite *via* solution casting. The hydrogen bonding interactions between chitosan and HNT caused uniform dispersion in the matrix. The mechanical properties such as tensile strength and Young's modulus of the nanomaterial were also enhanced with the HNT addition (Figure 5).

The 7.5wt.% HNT loading enhanced the tensile strength and tensile modulus to 54.2 MPa and 1240 MPa, respectively. Thus, the tensile strength and tensile modulus of the nanocomposite were enhanced by 134%and 65%, respectively, relative to the pure chitosan [29]. The 10 wt.% nanofiller loading enhanced the cytocompatibility of the chitosan/HNT nanocomposite. Hence, the chitosan/HNT bionanocomposites have found potential for the tissue engineering scaffolds.

Epoxy is a useful thermosetting polymer having unique properties. The epoxy and halloysite nanotube materials have been prepared using the 5 - 15wt.% nnaofiller. The incorporation of halloysite nanotubes has caused enhancement in the strength, modulus, toughness, and glass transition temperature (T_g) of the material [30, 31]. Tang et. al. [32] prepared the epoxy and 10wt.% HNT-based nanocomposite. The HNT was first unfold using phenylphosphonic acid (PPA). The fracture toughness K_{IC} of the unfold nanotube-based material was enhanced by 78.3%. The PPA treatment caused the better interaction and contact area between the epoxy and halloysite nanotubes. The T_g was evaluated from tanδ peak of dynamic mechanical analysis (DMA) curves. The T_g of the nanocomposite with the treated nanotubes was found higher than the untreated samples.

The poly(methyl methacrylate) (PMMA) is a hydrophobic thermoplastic polymer. It has been used in the structural materials owing to rigidity, transparency, and structural, opto-electronic, and biomedical properties [33]. However, the brittleness and poor impact resistance of the acrylic are the major drawbacks for the structural applicability [34]. PMMA has also been used as a matrix for HNT [35, 36]. Zhilin et. al. [37] prepared the PMMA/HNT nanocomposite nanofiber using the electrospinning technique. The materials have shown enhanced mechanical properties. The electrospinning also improved the orientation and dispersion of HNT in the PMMA nanofibers, so resulting in the enhanced thermal stability. Pal et. al. [38] designed the halloysite nanotubes and PMMA-based nanocomposites by using the solution route. The PMMA/HNT nanocomposites have shown high heat absorption capacity and mechanical properties, relative to the PMMA and carbon nanotube nanocomposite. The biocompatibility of HNT with PMMA render these materials suitable for the diverse bio-medical applications [39, 40].

THE APPLICATIONS OF POLYMER/HALLOYSITE NANOTUBE NANOCOMPOSITE

The polymer/halloysite nanotubes have been used in several technical applications such as corrosion protection, shape memory, flame protection, drug delivery, and tissue engineering. Corrosion of the metals is a serious technical difficulty. Several anti-corrosion agents have been proposed including halloysite and nanoclay additives [41]. The inhibitive performance of halloysite alone is not sufficient for the metal protection. The polymeric nanocompsoites with halloysite nanotube have, therefore, been used. Molaei et. al. [42] designed the chitosan/HNT nanocomposites for corrosion protection. The electrochemical corrosion tests proved the corrosion resistance of the chitosan/HNT for titanium substrate. Abdullayev et. al. [43] also studied the anticorrosive properties of the polymer/HNT coatings. The addition of the nanotubules into coatings enhanced the scratch resistance and anti-rusting of substrate. The shape memory polyurethanes have shown enhanced properties owing to the addition of halloysite nanotube [44]. Bouaziz et. al. [45] also prepared the shape memory polyurethane/HNT nanocomposites through the melt extrusion. The homogeneous distribution of nanofiller enhanced the recovery speed and cyclic shape memory of these nanocomposites. Zhu et. al. [46] prepared the polymer/HNT nanocomposite with high binding capacity (45.4mg/g) and specific identification for protein. The materials were found effective for the protein recognition and separation. The non-flammability behavior of the polymer/HNT nanocomposites have been studied. The green nanocomposites with halloysite nanotubes have been tested for the fire retardancy and thermal stability [47]. Attia et. al. [48] added the intumescent flame-retardants (melamine polyphosphate and pentaerythritol) to the acrylonitrile-butadiene-styrene and halloysite nanotube nanocomposites. The addition of the intumescent flame retardant enhanced the thermal stability and reduced the peak heat release rate to 56.2%. The nanocomposites also had reduced emission of CO and CO_2 gases. Rybiński et. al. [49] also studied the effect of halloysite nanotubes on the nano-flammability peroperties of the rubbers. The drug delivery propspects of the polymer/halloysite nanotube materials have been investigatd. The halloysite nanotube end capping with polymers may improve the drug-loading capabilities, release rate, and controlled duration of drug release [50, 51]. The polymer/halloysite nanotube nanocomposites have also been used in the tissue scaffolds and bone cement materials [9]. In the tissue engineering, HNT has been incorporated in alginate to improve the mechanical properties. The polymer/HNT has also increased the cell-attachment properties of the scaffolds [12, 52]. The tissue composite scaffolds have been studied for the mechanical, morphology, and physico-chemical properties [53].

CHALLENGES AND SUMMARY

The biggest advantage of the halloysite nanotubes is its natural occurance and the low cost. The surface modification of the nanotubes may improve interactions with the polymer matrices. However, the functionalization of HNT is difficult compared with the carbon nanofillers. Consequently, the HNT dispersion in the polymer matrix is a tiresome process. The morphological deviation in terms of the length, diameter, and wall thickness may also result in better performance and applications of polymer/HNT. The halloysite surface modification is also essential for drug delivery application. The tissue engineering scaffolds can also be better designed using the functional nanotube and polymers. For anti-corrosion applications, the polymer/HNT nanocomposites have the limited conductivity. The conductivity peoprties may be enhanced using the metal nanoadditives or other nanoparticles along with the HNT nanofiller. The non-flammability properties of the polymer/HNT can be increased using nanoclay additive with HNT. The future research is needed for thermo-responsive, light responsive, and pH sensitive polymer/HNT nanocomposite. Moreover, the halloysite nanotubes have not yet been used in electronics, energy, and aerospace applications. Therefore, the lots of research efforts are needed to replace carbon nanotube nanofiller in these technical areas.

This chapter highlights several stimulating features of the polymer and HNT-based nanostructured materials. The halloysite nanotubes have been used with polymers as a low cost nanofiller. The polymer/HNT nanomateraials have been developed using various techniques to attain superior thermal stability, non-flammability, strength, anti-corrosion, conductivity, shape memory, and biomedical properties. Consequrntly, these nanostructures have been employed for various functional and engineering applications. Future studies must focus the use of functional nanotubes and polymers to develop interfacial interactions in the nanomaterials. The better linked and dispersed polymer/HNT nanocomposites may lead to several potential hidden application areas of these materials for high efficiency applications.

REFERENCES

[1] Kausar, A., Review on polymer/halloysite nanotube nanocomposite. *Polymer-Plastics Technology and Engineering*, 2018. 57(6): p. 548 - 564.

[2] Du, M., B. Guo and D. Jia, Newly emerging applications of halloysite nanotubes: a review. *Polymer International*, 2010. 59(5): p. 574 - 582.

[3] Guimaraes, L. et al., Structural, electronic, and mechanical properties of single-walled halloysite nanotube models. *The Journal of Physical Chemistry C*, 2010. 114(26): p. 11358 - 11363.

[4] Pasbakhsh, P. et al. Halloysite nanotubes as a novel nanofiller for polymer nanocomposites. in *21st Australian Clay Minerals Conference*. 2010.

[5] Yuan, P., D. Tan and F. Annabi-Bergaya, Properties and applications of halloysite nanotubes: recent research advances and future prospects. *Applied Clay Science*, 2015. 112: p. 75 - 93.

[6] Liu, M. et al., Recent advance in research on halloysite nanotubes-polymer nanocomposite. *Progress in polymer science*, 2014. 39(8): p. 1498 - 1525.

[7] Cavallaro, G. et al., Films of halloysite nanotubes sandwiched between two layers of biopolymer: from the morphology to the dielectric, thermal, transparency, and wettability properties. *The Journal of Physical Chemistry C,* 2011. 115(42): p. 20491 - 20498.

[8] Wu, W. et al., Polylactide/halloysite nanotube nanocomposites: Thermal, mechanical properties, and foam processing. *Journal of Applied Polymer Science*, 2013. 130(1): p. 443 - 452.

[9] Rawtani, D. and Y. Agrawal, Multifarious applications of halloysite nanotubes: a review. *Rev. Adv. Mater. Sci.,* 2012. 30(3): p. 282 - 295.

[10] Zahidah, K. A. et al., Halloysite nanotubes as nanocontainer for smart coating application: A review. *Progress in Organic Coatings,* 2017. 111: p. 175 - 185.

[11] Yu, L. et al., Recent advances in halloysite nanotube derived composites for water treatment. *Environmental Science: Nano,* 2016. 3(1): p. 28 - 44.

[12] Liu, M. et al., Chitosan–halloysite nanotubes nanocomposite scaffolds for tissue engineering. *Journal of Materials Chemistry B,* 2013. 1(15): p. 2078 - 2089.

[13] Joussein, E. et al., *Halloysite clay minerals–a review.* 2005, De Gruyter.

[14] Weaver, C. E. and L. D. Pollard, *The chemistry of clay minerals,* Vol. 15. 2011: Elsevier.

[15] Vergaro, V. et al., Cytocompatibility and uptake of halloysite clay nanotubes. *Biomacromolecules,* 2010. 11(3): p. 820 - 826.

[16] Brindley, G., K. Robinson and D. MacEwan, The clay minerals halloysite and meta-halloysite. *Nature,* 1946. 157(3982): p. 225.

[17] Lvov, Y. et al., Halloysite clay nanotubes for loading and sustained release of functional compounds. *Advanced Materials,* 2016. 28(6): p. 1227 - 1250.

[18] Prashantha, K., M. F. Lacrampe and P. Krawczak, Highly dispersed polyamide-11/halloysite nanocomposites: Thermal, rheological, optical, dielectric, and mechanical properties. *Journal of Applied Polymer Science,* 2013. 130(1): p. 313 - 321.

[19] Guo, B. et al., Crystallization behavior of polyamide 6/halloysite nanotubes nanocomposites. *Thermochimica Acta,* 2009. 484(1-2): p. 48 - 56.

[20] Lecouvet, B. et al., Structure–property relationships in polyamide 12/halloysite nanotube nanocomposites. *Polymer Degradation and Stability,* 2011. 96(2): p. 226 - 235.

[21] Handge, U. A., K. Hedicke-Höchstötter and V. Altstädt, Composites of polyamide 6 and silicate nanotubes of the mineral halloysite:

influence of molecular weight on thermal, mechanical and rheological properties. *Polymer*, 2010. 51(12): p. 2690 - 2699.

[22] Zhao, M. and P. Liu, Halloysite nanotubes/polystyrene (HNTs/PS) nanocomposites via in situ bulk polymerization. *Journal of thermal analysis and calorimetry*, 2008. 94(1): p. 103 - 107.

[23] Liu, H. et al., Facile fabrication of polystyrene/halloysite nanotube microspheres with core–shell structure via Pickering suspension polymerization. *Polymer bulletin*, 2012. 69(7): p. 765 - 777.

[24] Lin, Y. et al., High-impact polystyrene/halloysite nanocomposites prepared by emulsion polymerization using sodium dodecyl sulfate as surfactant. *Journal of colloid and interface science*, 2011. 358(2): p. 423 - 429.

[25] Dutta, P. K., J. Dutta and V. Tripathi, *Chitin and chitosan: Chemistry, properties and applications.* 2004.

[26] Pillai, C., W. Paul and C. P. Sharma, Chitin and chitosan polymers: Chemistry, solubility and fiber formation. *Progress in polymer science*, 2009. 34(7): p. 641 - 678.

[27] Kumar, M. N. R., A review of chitin and chitosan applications. *Reactive and functional polymers*, 2000. 46(1): p. 1 - 27.

[28] Liu, M. et al., Chitosan/halloysite nanotubes bionanocomposites: structure, mechanical properties and biocompatibility. *International journal of biological macromolecules*, 2012. 51(4): p. 566 - 575.

[29] Okada, A. and A. Usuki, Twenty years of polymer-clay nanocomposites. *Macromolecular materials and Engineering*, 2006. 291(12): p. 1449 - 1476.

[30] Deng, S., J. Zhang and L. Ye, Halloysite–epoxy nanocomposites with improved particle dispersion through ball mill homogenisation and chemical treatments. *Composites Science and Technology*, 2009. 69(14): p. 2497 - 2505.

[31] Ye, Y. et al., High impact strength epoxy nanocomposites with natural nanotubes. *Polymer*, 2007. 48(21): p. 6426 - 6433.

[32] Tang, Y. et al., Effects of unfolded and intercalated halloysites on mechanical properties of halloysite–epoxy nanocomposites.

Composites Part A: *Applied Science and Manufacturing*, 2011. 42(4): p. 345 - 354.

[33] Ali, U., K. J. B. A. Karim and N. A. Buang, A review of the properties and applications of poly (methyl methacrylate)(PMMA). *Polymer Reviews*, 2015. 55(4): p. 678 - 705.

[34] Gad, M. M. et al., PMMA denture base material enhancement: a review of fiber, filler, and nanofiller addition. *International journal of nanomedicine,* 2017. 12: p. 3801.

[35] Wang, L.-P. et al., Synthesis of poly (methyl methacrylate)-b-poly (N-isopropylacrylamide)(PMMA-b-PNIPAM) amphiphilic diblock copolymer brushes on halloysite substrate via reverse ATRP. *Reactive and Functional Polymers*, 2008. 68(2): p. 649 - 655.

[36] Wei, W. et al., Clay nanotube/poly (methyl methacrylate) bone cement composites with sustained antibiotic release. *Macromolecular materials and engineering,* 2012. 297(7): p. 645 - 653.

[37] Zhilin, C., Q. Xixi and Q. Dunzhong, Electrospinning preparation and mechanical properties of polymethyl methacrylate (PMMA)/halloysite nanotubes (HNTs) composite nanofibers. *China Petrol. Process. Petrochem. Technol.,* 2016. 18: p. 52 - 56.

[38] Pal, K., Effect of different nanofillers on mechanical and dynamic behavior of PMMA based nanocomposites. *Composites Communications,* 2016. 1: p. 25 - 28.

[39] Chen, S., Z. Yang and F. Wang, Investigation on the Properties of PMMA/Reactive Halloysite Nanocomposites Based on Halloysite with Double Bonds. *Polymers,* 2018. 10(8): p. 919.

[40] Vuluga, Z. et al., Morphological and Tribological Properties of PMMA/Halloysite Nanocomposites. *Polymers,* 2018. 10(8): p. 816.

[41] Lvov, Y. M. et al., Halloysite clay nanotubes for controlled release of protective agents. *ACS nano,* 2008. 2(5): p. 814 - 820.

[42] Molaei, A. et al., Structure, apatite inducing ability, and corrosion behavior of chitosan/halloysite nanotube coatings prepared by electrophoretic deposition on titanium substrate. *Materials Science and Engineering: C,* 2016. 59: p. 740 - 747.

[43] Abdullayev, E., D. Shchukin and Y. Lvov, Halloysite clay nanotubes as a reservoir for corrosion inhibitors and template for layer-by-layer encapsulation. *Mater. Sci. Eng.,* 2008. 99: p. 331 - 332.

[44] Ismail, H. et al., The effect of halloysite nanotubes as a novel nanofiller on curing behaviour, mechanical and microstructural properties of ethylene propylene diene monomer (EPDM) nanocomposites. *Polymer-Plastics Technology and Engineering,* 2009. 48(3): p. 313 - 323.

[45] Bouaziz, R., K. Prashantha and F. Roger, Thermomechanical modeling of halloysite nanotube-filled shape memory polymer nanocomposites. *Mechanics of Advanced Materials and Structures,* 2018: p. 1 - 9.

[46] Zhu, X. et al., Fabrication and evaluation of protein imprinted polymer based on magnetic halloysite nanotubes. *RSC Advances,* 2015. 5(81): p. 66147 - 66154.

[47] Nakamura, R. et al., Effect of halloysite nanotubes on mechanical properties and flammability of soy protein based green composites. *Fire and Materials*, 2013. 37(1): p. 75 - 90.

[48] Attia, N. F. et al., Flame-retardant materials: synergistic effect of halloysite nanotubes on the flammability properties of acrylonitrile–butadiene–styrene composites. *Polymer International*, 2014. 63(7): p. 1168 - 1173.

[49] Rybiński, P. et al., Thermal properties and flammability of nanocomposites based on diene rubbers and naturally occurring and activated halloysite nanotubes. *Journal of thermal analysis and calorimetry,* 2011. 107(3): p. 1243 - 1249.

[50] Lvov, Y. M., M. M. DeVilliers and R. F. Fakhrullin, The application of halloysite tubule nanoclay in drug delivery. *Expert opinion on drug delivery,* 2016. 13(7): p. 977 - 986.

[51] Ghebaur, A., S. A. Garea and H. Iovu, New polymer–halloysite hybrid materials—potential controlled drug release system. *International journal of pharmaceutics*, 2012. 436(1-2): p. 568 - 573.

[52] Fakhrullin, R. F. and Y. M. Lvov, *Halloysite clay nanotubes for tissue engineering.* 2016, Future Medicine.

[53] Liu, M. et al., In vitro evaluation of alginate/halloysite nanotube composite scaffolds for tissue engineering. *Materials Science and Engineering: C,* 2015. 49: p. 700 - 712.

In: Halloysite
Editor: Herbert A. Eckart

ISBN: 978-1-53616-812-9
© 2020 Nova Science Publishers, Inc.

Chapter 4

RECENT ADVANCES IN HALLOYSITE NANOTUBES APPLICATIONS

Dorel Florea, Mihai Cosmin Corobea and Zina Vuluga[*]

National Institute for Research and Development in Chemistry and Petrochemistry -ICECHIM, Bucharest, Romania

ABSTRACT

Halloysite nanotubes (HNT) were proven as versatile structures because of their outstanding properties such as tubular or "scroll-like" morphology, high mechanical strength and thermal stability. Moreover, their crystalline structure and low hydroxyl density on the surface helps them to disperse relative easily in various polymer matrices.

They are considered as cost accessible and can be found in both natural and synthetic form. HNT general availability, recommended them already for large number of applications.

The present chapter summarizes the latest research concerning the HNT applications, in different fields and topics like biomedicine and pharmacotherapy, food packaging, agriculture, water treatment, catalysis, or antifouling. A special attention was given to HNT as reinforcement for polymer composites (based on different polymer matrix like PA6, PA11,

---

[*] Corresponding Author's E-mail: zvuluga@yahoo.com.

PMMA, PLA, PS, PP, PVA, EVA, PHB, PEG, thermoplastic starch, or CMC).

This chapter covers research on the use of Halloysite for the last 4 years.

LIST OF ABBREVIATIONS

Thermoplastic Polymers

ABS	Acrylonitrile Butadiene Styrene
COC	Cyclic Olefin Copolymer
EVA	Ethylene-Vinyl Acetate
HDPE	High Density Polyethylene
HIPS	High impact Polystyrene
iPP	isotactic Polypropylene
LDPE	Low Density Polyethylene
LLDPE	Linear Low-Density Polyethylene
PA	Polyamide
PA6	Polyamide 6
PA11	Polyamide 11
PBAT	Poly(Butylenes Adipate-co-Terephthalate)
PBS	Poly(Butylenes Succinate)
PC	Polycarbonate
PCL	Poly(ε-caprolactone)
PEI	Polyetherimide
PHBV	Poly(3-hydroxybutyrate-co-3-hydroxyvalerate)
PLA	Poly-lactic acid
PMMA	Polymethyl methacrylate
PP	Polypropylene
PS	Polystyrene
PTFE	Polytetrafluoroethylene
PVA	Polyvinyl alcohol
UHMWPE	Ultrahigh molecular weight Polyethylene

Natural Polymers

Carrageenan	Sulphated polysaccharides
CMC	Carboxymethyl cellulose

Compatibilising Agents

PE-g-MA	Polyethylene graft Maleic Anhydride
PP-g-MA	Polypropylene graft Maleic Anhydride
SEBS-g-MA	Styrene-ethylene/butylene-styrene graft maleic anhydride
ENR-50	(Epoxidized natural rubber, 50% mol of epoxidation)
P(S-co-MAPC$_1$(OH)$_2$)	Styrene/(methacryloyloxy)methyl phosphonic acid copolymer
SMA	Styrene-maleic anhydride Copolymer

Fillers

GF	Glass Fibres
HNTs	Halloysite Nanotubes
MCC	Microcrystalline Cellulose
MFC	Microfibrillar Composite

Flame Retardants

IFR	Intumescent flame retardant (microencapsulated ammonium polyphosphate/pentaerythritol phosphate 2:1)
FR	additive based on metal phosphinate

Irgafos 168	tris(2,4-di-tert-butylphenyl) Phosphate
DOPO	9,10-dihydro-9-oxa-10-phosphaphenantrene-10-oxide
BDP	Bisphenol A bis(diphenyl phosphate)

Coupling Agents

KH550	γ-aminopropyltriethoxysilane
KH570	3-Methacryloxypropyltrimethoxy silane
APTES, ASP	3-aminopropyltriethoxysilane
APTS	(3-Aminopropyl)-trimethoxysilane
DAS	N- [3-(trimethoxysilyl)propyl] ethylendiamine, N-(2-Aminoethyl-3-aminopropyl)trimethoxysilane
GLYMO	3-glycidyloxypropyl trimethoxysilane
MPS	3-trimethoxysilylpropyl methacrylate
EPB	2,2-(1,2-ethene diyldi-4,1-phenylene) Bisbenzoxazole

Antioxidants

| Irganox 1010 | 3-(3,5-di-tert-butyl-4- hydroxyphenyl) propionate |
| TNPP | tris(nonylphenyl) phosphate |

Anorganic Substances

| NaOH | Natrium hydroxide |
| ZrO_2 | Zirconia dioxide |

Thermoset Polymer

    PU   Polyurethane

Polyether Compound

    PEG   Poly(Ethylene Glycol)

Polysaccharide

    XG   xanthan gum

Elastomers

    NBR   Nitrile Butadiene Rubber copolymer

Characteristics of Materials

DR	Draw Ratio
E_a	Activation Energy
ESR	Equilibrium Swelling Ratio
IT	Ignition Temperature
LOI	Limiting Oxygen Index
MLM	Mechanical Loss Magnitude
OOT	Oxidation Onset Temperature
OTR	Oxygen Transmission Rate
PHRR	Peak of Heat Release Rate
T_c	Crystallisation Temperature
T_{cc}	Cold Crystallisation Temperature
T_g	Glass Transition Temperature

Tm	Melting Temperature
$T_{50\%}$	Temperature of 50% weight loss
TSR	Total Smoke Release
TTI	Time to Ignition
WVTR	Moisture Vapour Transmission Rate
Xc	Degree of crystallinity

Antibacteria Agent

TCN	Triclosan 2,4,4-Trichloro-2 hydroxydiphenyl ether

Anionic Surfactant

SDS	Sodium dodecyl sulfate

Crosslinking Agent

DCP	Dicumyl peroxide

Impact Modifier or Lubricant

EBS	N,N'-ethylenebis(stearamide)

Drug

BG	Triazole dye brilliant green
DOX	Doxorubicin hydrochloride

Catalyst

    CuI@HNTs -   2 N - Sal (salicylate)

Vitamins

    FA   Folic acid
    Biotin   Vitamin H

Antifoulant

    TCPM   N-(2,4,6-Trichlorophenyl) Maleimide

Dendrimer

    PAMAM G(1/2)   polyamidoaminedendrimer generation (1/2= 8/16 surface groups)

Various

    FITC   Fluorescein isothiocyanate
    Binase   The ribonuclease from *Bacillus pumilus*

Methods of Characterisation

    WAXD   Wide-Angle X-Ray Diffraction
    DSC   Differential scanning calorimetry

DMA   Dynamic mechanical analysis
TGA   Thermogravimetric analysis
WVP   Water Vapou r Permeability

INTRODUCTION

For over a decade, a growing interest has focused on the use of halloysite nanotubes as nanocontainers for adsorption of organic molecules, herbicide, antifouling, flame retardant, filler for reinforcement of thermoplastic polymer composites, in environment protection, for encapsulating and controlled release of drugs.

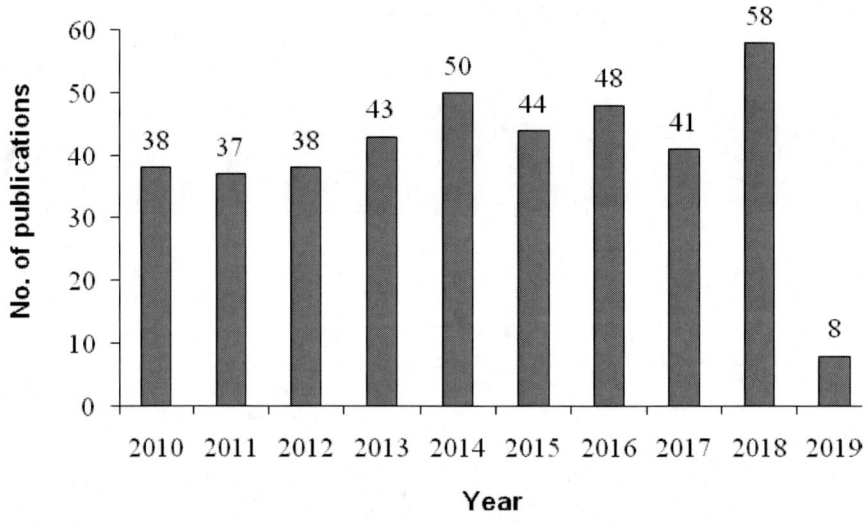

Figure 1. Scientific papers per year based on the words "polymer halloysite nanocomposites" (data analysis on 5th March 2019).

An examination of open publications about "polymer halloysite nanocomposites" in the year's interval 2010-2019 has shown a constant interest in this topic. The information was collected based on a PubMed search.

To avoid unnecessary redundancy with other reviews or book chapters published in the literature, herein, we present only relatively recent (four years) advances on the applications of the halloysite nanotubes are reviewed.

HALLOYSITE NANOTUBES IN WIDE APPLICATION DOMAIN

Halloysite Nanotubes as Reinforcement for Thermoplastic Polymer Composites

HNTs as Reinforcement for Biodegradable-Based Polymer Composites

The biodegradable polymers can (in manny cases) be degraded or dissolved into the environment without high impact on pollution. In some cases, they are biocompatible homopolymer with high crystallinity, having poor mechanical properties. Their applications are various from medical, i.e., drug release, agriculture up to packaging fields. The polysaccharides are a principal class of biodegradable polymers, and among them of interest are the cellulose, starch and chitosan. In addition, as points of interest can be mentioned PHBV, PBS, and PVA.

Cellulose consists of glucose units that form the main part of the cell walls of plants. Starch is the main storage form of carbohydrates in plants (Miller and Keane, 2003). He is non-toxic, of low-cost, has low mechanical properties and for meeting the application demands blending with other materials such as polymers (chitosan (Shen et al. 2010), PVA (Tian et al. 2017) or fillers (clays (Zeppa et al. 2009), titanium dioxide)) is a necessity.

Chitosan, the partly deacetylated form of chitin, is biocompatible, has antimicrobial activity, can absorb heavy metal ions, and is used as drug carrier molecules due to its cationic nature.

However, chitosan has the drawbacks of poor gel strength, as well as low stability and mechanical properties. For overcome these, various methods have been utilised (i.e., using crosslinking agents, incorporation of nanofillers, etc.).

PHBV is a biodegradable, non-toxic, biocompatible polymer used in specialty packaging, and medicine (i.e., medical implants and controlled release of drugs). However, PHBV is brittle, has low thermal stability, low elongation at break and impact resistance (Avella et al. 2000).

PBS is biodegradable aliphatic polyester showing potential commercial applications due to its good processability, and better thermomechanical property. However, its further applications are limited because of its poor impact strength, low melt viscosity, and insufficient stiffness. The incorporation of various inorganic fillers such as clay, silica, etc. has been proven as an effective approach to enhance the performance of PHBV, PBS, or their blends (Wu et al. 2014, Kennouche et al. 2016).

PVA is a water-soluble synthetic polymer, widely used in biomedical applications due to its favourable properties such as biocompatibility, nontoxicity, and swelling. Also, PVA-based composites are used in packaging industry, pharmaceutical, and in tissue engineering (Gaaz et al. 2015). For applications where insolubility in water is necessary, as well as for the enhancing of mechanical properties the addition of fillers such as HNT is very useful.

PBAT is a biodegradable random copolymer having physical properties such as high flexibility and toughness, but low elastic modulus and stiffness. The drawback of PBAT consists on slow crystallisation rate, low degree of crystallinity, low tensile stress, which limits its application in medical engineering or structure materials (Fukushima et al. 2012). Incorporation of nanofillers into PBAT can regulate its crystallisation behaviour, increase the hardness, improve the mechanical properties, and broaden its application.

Compatibilising with Grafted Polymer

PHBV/PBS blends have been prepared by Kennouche et al. (2016), as well as a MA-grafted PHBV compatibiliser. The effect of MA-g-PHBV and HNTs, on the morphological and thermal properties of PHBV/PBS blends was studied. The dispersion of HNTs in the blend has the effect of improving the fire resistance and decreasing the thermal stability of the blend. Microscopic studies have shown that HNTs are localised in PBS. PHBV-g-MA chains localise at the PHBV/PBS interface and reduce interfacial tension, without significantly affecting the thermal properties of the blend. In the ternary nanocomposites, the authors hypothesised the interaction between PHBV-g-MA and HNTs and the formation of aggregates that impede the diffusion of PHBV-g-MA at the interface. This phenomenon has the effect of changing the thermal properties, but not as significant as in the case of adding only HNTs.

Gaaz and Hussein (2017) were produced HNTs based nanocomposites by blending customised HNTs dispersion with PVA. HNTs are customised using many techniques and the obtaining membrane-as biodegradable HNTs-PVA nanocomposites are then fabricated. These membrane-like nanocomposites are characterised by smooth surfaces while their thermal and tensile properties are excellent. PVA and its HNTs-PVA nanocomposites were crosslinked using MA. They asserted that an important achievement for PVA-crosslinking is that HNTs-PVA composite is characterised as water insoluble. Using MA crosslinked HNTs-PVA composites shows much better improvement of the thermal and mechanical properties.

Pre-Treatment with Coupling Agent

In a study done by Li et al. (2018), the effect of interfacial interaction between HNT/surface modified HNT and PBAT on the non-isothermal crystallisation behaviour of PBAT was investigated. HNT and HNT modified with APTES acted as nucleation agent for PBAT, the nanocomposites having a lower crystallisation rate than the calcinated-HNT/PBAT and APTES modified calcinated-HNT/PBAT nanocomposites. The APTES modified calcinated-HNT slowed the transport of PBAT molecules chains and decreased the crystallinity of PBAT. Also, the mechanical properties and thermal behaviour were increased.

Garcia-Garcia et al. (2018) studied the use of caffeic acid both as a surfactant for reducing the hydrophilic properties of HNT and as an antioxidant for improving thermal stability. The effect of HNT modified with caffeic acid was studied compared to the effect of HNT modified with GLYMO on the properties of the PHB/PCL blends. The authors concluded that by silanisation the hydrophobicity of HNT increased, which was reflected in the increased mechanical properties of the PHB/PCL blends (partially compatibilised with DCP by reactive extrusion).

Sengel et al. (2017) reported the preparation of super porous composite cryogels using CMC and HNT. HNT was functionalised with various modification agents (i.e., APTES, PEI, tris(2-aminoethyl)amine, (3-chloro-2-hydroxypropyl)trimethylammonium chloride, epichlorohydrin, diethylenetriamine, and taurine) and then used in HNT/CMC composites. Pore sizes, softness / hardness and water absorption capacity of cryogel composites were different depending on the amount of crosslinker and HNT used.

No Treatment, or Surfactant Treatment

Suppiah et al. (2019) studied the CMC/HNT nanocomposites for biodegradable films. The limitations of these bio-nanocomposites such as sensitivity toward moisture, tendency to form agglomerates were reduced by the functionalisation of HNT using SDS, for achieving a certain extent of stiffness, strength and barrier properties. They reported that treated HNT/CMC bio-nanocomposites showed enhanced mechanical properties, improved hydrophobicity, and thermal stability.

In a study done by Suppiah et al. (2017) the effect of HNT filler content on tensile properties and morphology of CMC/HNT bionanocomposite films were investigated. They reported the increasing of mechanical properties of CMC/HNT bio-nanocomposite films.

Huang et al. (2017) prepared chitosan composite hydrogels with HNTs by heat treating lithium hydroxide/potassium hydroxide/urea composite solution and regenerating it in ethanol solution. They reported a significant increase in the mechanical properties and anti-deformation properties of the chitosan hydrogels. The composite hydrogels showed an excellent biocompatibility. Also, DOX loading ability of composite hydrogels is increased with the increase of HNTs content.

Electrostatic assembly process of HNTs porous microspheres and its modification as drug carrier for aspirin was reported by Li et al. (2016). They showed that HNT/chitosan/Aspirin nanocomposites had the lower release amount of Aspirin in the simulated gastric fluid, whilst more rapid release rate and higher release amount in the simulated intestinal fluid, reducing side effects to stomach.

Madhusudana et al. (2018a) studied the effects of HNT morphology and polyelectrolyte complexation of chitosan and xanthan gum on the structural interactions (with both lumen and outer surface of HNTs), swelling behaviour, crystalline nature, microstructure, and mechanical properties of nanocomposite hydrogels for biomedical applications.

Choo et al. (2016) concentrated on production of chitosan/HNT nanocomposite beads through ultrasonic-assisted extrusion-dripping technique. They studied the effects of ultrasonic amplitude and HNT loading fractions on the characteristics and reported the adsorption performance of the chitosan/HNT beads, for nanocomposite adsorbents with desired adsorption capacity for copper ion removal at low concentrations.

Madhusudana et al. (2018b) reported a facile in-situ method for encapsulation of curcumin and Au nanoparticles into lumen as well as the surface cage of HNT followed by coating with the chitosan as biopolymer for cancer drug delivery. The hybrid nanoparticles showed pH-controlled drug delivery behaviour with greater cytotoxicity on cancer cells in acidic conditions.

A method to fabricate enzymatic membranes using HNTs, chitosan and enzymes with antimicrobial activities – lipase and lysozyme was reported by Sun et al. (2017). Started with lipase loading into the lumen of the HNT (and optimising her), then the enzyme-HNT complex was mixed with chitosan for obtaining the membrane. The obtained HNT/enzyme chitosan membranes showed good catalytic lipid decomposition and stability. Then, the cationic lysozyme was adsorbed onto the HNTs outer surface, thus increasing the antifouling capability of the dual enzyme composite (Figure 2).

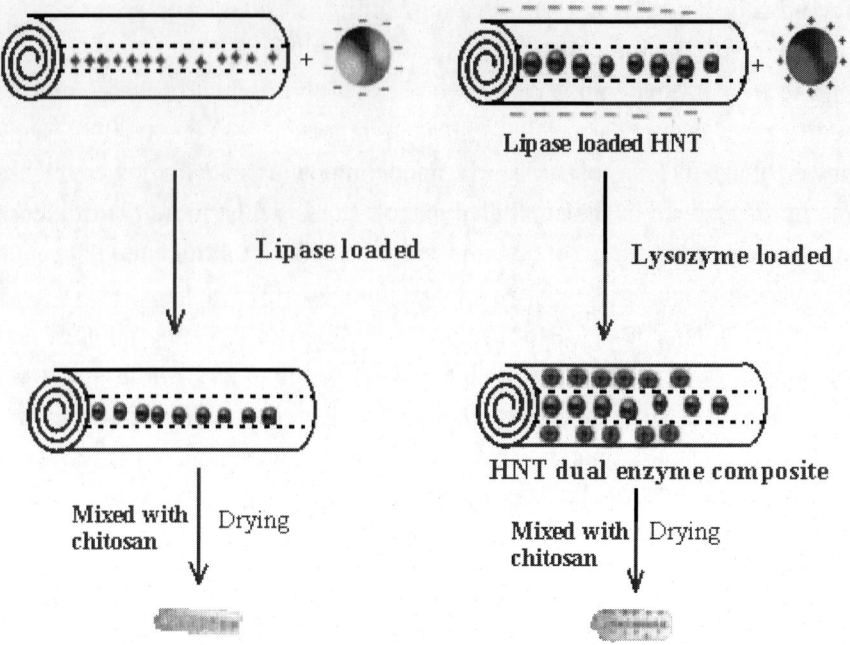

Figure 2. Loading the anionic lipase into the tube lumen, then the cationic lysozyme onto the surface of the negative tube, and the membrane preparation (Adapted from Sun et al. 2017).

A short review is provided by Gaaz et al. (2015) for the synthesis, properties and applications of PVA-HNTs and their nanocomposites. They emphasised the individualised HNT addition and crosslinking of PVA for various biomedical applications. They concluded that PVA-HNT nanocomposites can be used in general clinical operations such as cartilage replacement or cartilage transplantation. Moreover, these nanocomposites can be used in eye contact lenses, eye drops, and drug delivery systems to target tissue with abnormal growth rates.

Cheng et al. (2017a) studied the structure and mechanical properties of the PVA/HNT composite nanofibers prepared by electrospinning. They revealed that the tensile modulus and tensile strength at 15 wt. % of HNT content exhibited a good performance.

HNTs incorporated as second nanofiller into PVA matrix in order to assess their ability to provide a better dispersion of cellulose nanocrystals CNC and to confer better performance to the reinforced PVA was studied by Aloui et al. (2016). They evaluated the effect of various amounts of HNTs and CNC on the properties of the developed films. They revealed that HNT and CNC incorporation have improved the mechanical and thermal properties of PVA matrix, attesting for the good interfacial adhesion between the reinforcement phase and the matrix. As well as, they showed a significant reduction of water sorption and increase in the water barrier properties of PVA matrix, due to a synergistic effect between CNC and HNTs, based on specific interactions between their OH groups.

For improving the flexibility of bio-nanocomposites based on PVA/starch and enhancing the processability, Abdullah and Dong (2018) studied the embedding of HNTs and glycerol in this system. They reported an enhancement of Young's modulus, a limited improvement of tensile strength and increased Tg and Xc by addition of HNTs from 0.25 to 1 wt. %.

Oliyaei et al. (2019) studied the improvement of adsorption capacity of porous starch by incorporation of HNTs with different ethanol/water ratio because of the ability of ethanol for maintenance of pores. They reported that the degree of porosity could be modulated by selecting the type of drying and solvent ratio or combination with HNT, thus creating new pathways for innovative active packaging or medical devices applications.

Nanocomposite films of HNT and carrageenan were prepared by Wahab et al. (2017), to investigate their mechanical properties. They reported the improvement of the strength and modulus for an optimum loading of 3 % HNT.

Table 1 summarise the main properties of HNTs/biodegradable polymer nanocomposites in relation with the principal obtaining routes.

Table 1. Various HNTs/Polymer nanocomposites

Nanocomposites	HNTs modifier	HNT (%)	Mixing method	Properties variations (%)(compared to neat resin)	Ref.
HNT/PHBV/PBS	PHBV-g-MA	5	Melt blending		Kennouche et al. 2016
HNTs/PVA (crosslinked MA nano-composites)	-		Blending and casting		Gaaz, and Hussein 2017
HNT/PBAT	APTES (3-aminopropyltriethoxy-silane)	2	Solution casting		Li et al. 2018
HNT/PHB/PCL/ DCP (1 phr) (3 / 1)	(3-glycidyloxypropyl trimethoxysilane) (GLYMO), caffeic acid (CA)	3	Melt blending	3HNT/PHB/PCL/DCP -tensile strength: -20 -elongation at break: -1.5 -tensile modulus: -26.7 -flexural strength: -17.5 3HNT-Silan/PHB/PCL/DCP -tensile strength: -10 -elongation at break: -7.7 -tensile modulus: -15 -flexural strength: -11.5 3HNT-CA/PHB/PCL/DCP -tensile strength: -9.5 -elongation at break: -15.7 -tensile modulus: -5.3 -flexural strength: -17.8	Garcia-Garcia et al. 2018

Nanocomposites	HNTs modifier	HNT (%)	Mixing method	Properties variations (%)(compared to neat resin)	Ref.
HNT/CMC	(3-aminopropyl)triethoxysilane; (3-chloro-2-hydroxypropyl)trimethylammonium chloride; polyethylenimine; epichlorohydrin; diethylenetriamine; taurine; tris(2-aminoethyl)amine	-	Cryogelation		Sengel et al. 2017
HNT/CMC-SDS	SDS (3 %)	5, 10, 15, 20	Solution casting	5HNT/CMC-SDS -moisture content: -29.4 -tensile strength : +26.7 -elongation at break: +31.7 -Young's modulus: +18.3 10HNT/CMC-SDS -moisture content: -40.1 -tensile strength : +46.9 -elongation at break: +51.2 -Young's modulus: +38.4 15HNT/CMC-SDS -moisture content: -53.6 -tensile strength : +36.2 -elongation at break: +84.6 -Young's modulus: +55 20HNT/CMC-SDS -moisture content: -66.1 -tensile strength : +14.8 -elongation at break: +96.8 -Young's modulus: +65	Suppiah et al. 2019

Table 1. (Continued)

Nanocomposites	HNTs modifier	HNT (%)	Mixing method	Properties variations (%)(compared to neat resin)	Ref.
HNTs/CMC	-	0 - 20	Solution casting	5HNT/CMC -tensile strength: +3.3 -elongation at break: +20 -Young's modulus:+2.6 10HNT/CMC -tensile strength: +33.7 -elongation at break: +29.5 -Young's modulus:+21 15HNT/CMC -tensile strength: +15.2 -elongation at break: +56 -Young's modulus:+32 20HNT/CMC -tensile strength: +8.2 -elongation at break: +82.8 -Young's modulus:+48	Suppiah et al. 2017
HNT/Chitosan	-	33.3, 50, 66.6	Solution casting		Huang et al. 2017
HNT/Chitosan		1	Micro-emulsion		Li et al. 2016

Nanocomposites	HNT's modifier	HNT (%)	Mixing method	Properties variations (%)(compared to neat resin)	Ref.
HNT/Chitosan	Xanthan gum	2.5 - 10	Solution	2.5HNT-XG /Chitosan -ESR: -5.3 -compressive strength: +8.9 -stiffness: +7.2 5HNT-XG /Chitosan -ESR: -18.4 -compressive strength: +33.3 -stiffness: +28.2 7.5HNT-XG /Chitosan -ESR: -22.1 -compressive strength: +54 -stiffness: +52 10HNT-XG /Chitosan -ESR: -26.3 -compressive strength: +102.7 -stiffness: +117.8	Madhusudana Rao et al. 2018a
HNT/Chitosan	-	25, 50, 75	Solution mixing		Choo et al. 2016
HNT/Curcumin-Au/Chitosan	Curcumin-Au	-	Solution		Madhusudana Rao et al. 2018b
HNT-lipase/Chitosan	Lysozyme (5:1), lipase (2:1)	10	Solution	HNT-lipase/Chitosan -relative activity (80°C): +2.6	Sun et al. 2017
HNTs/PVA	-	0 - 12	Dispersion electro-spinning		Gaaz et al. 2015

Table 1. (Continued)

Nanocomposites	HNTs modifier	HNT (%)	Mixing method	Properties variations (%)(compared to neat resin)	Ref.
HNT/PVA	-	5 - 25	Electro-spinning	5HNT/PVA -tensile strength: +65.6 -elongation at break: +19.6 -tensile modulus: +10.2 10HNT/PVA -tensile strength: +115.3 -elongation at break: +11.9 -tensile modulus: +10.2 15HNT/PVA -tensile strength: +214.5 -elongation at break: -5.4 -tensile modulus: +42.4 20HNT/PVA -tensile strength: +169.6 -elongation at break: -15.8 -tensile modulus: +72.4 25HNT/PVA -tensile strength: +138.4 -elongation at break: -29 -tensile modulus: +32.7	Cheng et al. 2017a

Nanocomposites	HNTs modifier	HNT (%)	Mixing method	Properies variations (%)(compared to neat resin)	Ref.
HNT/PVA/CNC (from MCC) (CNC 1, 5 wt.%)	-	1, 3	Solvent casting	1HNTs/PVA/1CNC -tensile strength: +69.7 -elongation atbreak: +77.3 -water vapour permeability: -15.2 1HNTs/PVA/5CNC -tensile strength: +68.2 -elongation at break:+79.3 -water vapour permeability: -19.7 3HNTs/PVA/1CNC -tensile strength: +67.4 -elongation at break:+100 -water vapour permeability: - 3 3HNTs/PVA/5CNC -tensile strength: +72.7 -elongation at break:+100 -water vapour permeability: -39.4	Aloui et al. 2016
HNT/PVA/ Starch/Glycerol	-	0.25, 0.5, 1,3,5	Solution casting	0.25HNT/PVA/St/Glyc -tensile strength: +20 -elongation at break: +5.9 -Young's modulus: +30 0.5HNT/PVA/St/Glyc -tensile strength: +3 -elongation at break: -39.6 -Young's modulus: +100 1HNT/PVA/St/Glyc -tensile strength: -11 -elongation at break: -36.3 -Young's modulus: +141.7 3HNT/PVA/St/Glyc	Abdullah and Dong 2018

Table 1. (Continued)

Nanocomposites	HNTs modifier	HNT (%)	Mixing method	Properties variations (%)(compared to neat resin)	Ref.
HNTs/Starch	-	0.1 - 0.5	Solvent exchange	-tensile strength: -22 -elongation at break: -42.3 -Young's modulus: +60.4 5HNT/PVA/St/Glyc -tensile strength: -38 -elongation at break: -43.2 -Young's modulus: +39.6	Oliyaei et al. 2019
HNT/Carrageenan	-	1 - 4	Solution blending	HNT/Carrageenan -tensile strength: +18.7 -elongation at break: -4.4 2HNT/Carrageenan -tensile strength: +25 -elongation at break: -6.8 3HNT/Carrageenan -tensile strength: +39.6 -elongation at break: -17.9 4HNT/Carrageenan -tensile strength: +10.5 -elongation at break: -6.2	Wahab et al. 2017

HNTs as Reinforcement for Polyamide-Based Composites

Polymer matrix composite is the material which consists of a polymer matrix combined with filler dispersed phase. The reinforcement of polymers by the addition of nanofillers such as clay minerals or halloysite nanotubes has attracted much attention in polymer matrix composite since many years. Apart from the combination of the properties of polymer matrix and fillers and processing like conventional plastics, the small dimensions of the nanofillers lead to specific properties such as transparency (Handge et al. 2010). Moreover, heat resistance, flame retardance, and wear properties of polymers can be enhanced by adding nanofillers.

The polyamide is a thermoplastic polymeric substance having recurring amide groups along the polymeric chain. PA 6 and PA 11 are between the polyamides which incorporate which can be obtained starting from one single monomer. PA11 has a good oil resistance and is less hydrophilic than PA6. Both polyamides had been extensively applied in many fields because they belong to a class of polymers close to high performance polymers class (based on properties/price ratio).

Compatibilising with Grafted Polymer and Pre-Treatment with Coupling Agent

However, just like all polyamides, PA11 is quite brittle and notch sensitive at high strain rates or at low temperatures (Rinawa et al. 2014). For counterbalancing this embrittlement of PA11, incorporation of a suitable elastomeric phase has been used in many cases. For enhancing compatibility with PA, an olefinic elastomer (like SEBS) must be grafted with maleic anhydride (MA), and then the MA group will react with the amine end-group of the PA, thus improving the dispersion of elastomer (Borggreve and Gaymans 1989).

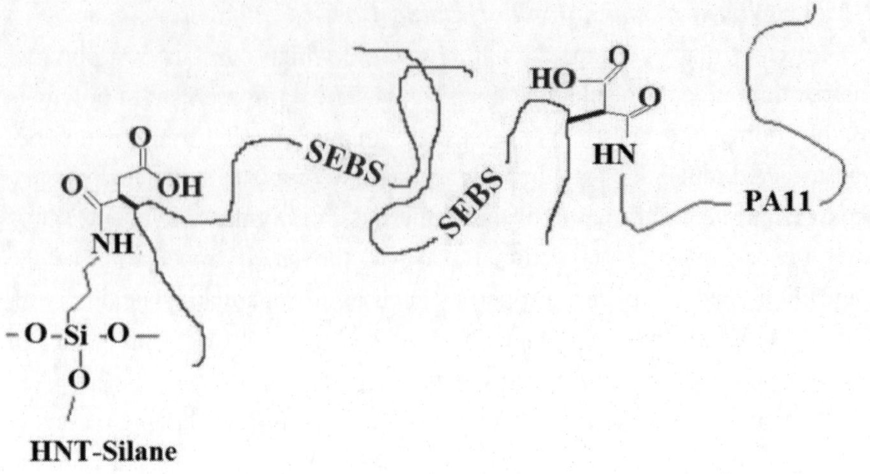

Figure 3. Schematic representation of the entanglements between grafted SEBS of HNT-silan and the SEBS-g-MA in their PA11/SEBS-g-MA blend (Adapted from Sahnoune et al. 2017 a).

Some researches (Sahnoune et al. 2017a) focused on enhancement of the toughness and the thermal properties, as well as on understanding morphologies of PA11/SEBS-g-MA blends brought by the functionalised HNT with SEBS chains on its outer surface (Figure 3). Incorporation of functionalised HNT (with APTES) leads to an improvement of the thermal properties, crystallisation of PA11 remaining unchanged. Also, mechanical performance, mainly toughness, was improved due to a good stress transfer from the polymer matrix to the modified HNT agglomerates surrounded by SEBS-g-MA. Other investigations (Sahnoune et al. 2017b) focused on the influence of HNT modification, via grafting of PS-co-MAPC$_1$(OH)$_2$) copolymer chains at the inner surface, on the morphology of PS/PA11 blends (80/20 and 60/40 wt. %). After the modification, HNT changed its location, migrating from the core of the PA11 phase to the polymer matrix interface. Also, thermal stability of the polymer matrix is enhanced after HNT modification mainly in 60/40 blend, and HNT act as a nucleating agent.

Paran et al. (2017) studied the effect of HNTs modification on the morphology and mechanical properties of the PA6/NBR (60/40 w/w). They concluded that introduction of silane modified HNT in PA6/NBR nanocomposites induces the reduction of the size of NBR droplets in thermoplastic elastomer nanocomposites. The Young's modulus was increased up to 45% and 75% respectively by introducing of pure and modified HNTs into the PA6 phase. The results of WAXD showed the transition of α-phase crystal structure of PA6 phase to the γ-phase structure due to the introduction of pure and organosilane modified HNTs.

No Treatment

In a study done by Cheng et al. (2018a), the HNTs/disc-shaped diatomite mixture was used to study the synergistic reinforcing effect of the filler in PA6 matrix. The mechanical performance of the composites was investigated as well as the structure of HNTs/diatomite mixture filler by XRD and SEM. The results showed that the optimum reinforcing performance was at the HNTs/diatomite mixture proportion of 5:1, suggesting that the above mixture filler could have a synergistic effect on reinforcing polymer.

Boonkongkaew et al. (2018) studied the combined use of HNT and organophosphorus flame retardant BDP for improving the flame retardancy of PA6. A certain amount of BDP (2 wt. %) was encapsulated inside the HNT lumen before compounding the BDP-loaded HNT with the PA6, for reducing the negative influences of BDP on the mechanical properties and thermal stability of the PA6 nanocomposites. They obtained a V-0 rating in UL-94 test with a decrease in the PHRR and THR characteristics.

In another study Boonkongkaew et al. (2017) examined the structural effects of phosphite antioxidants on the thermo-oxidative ageing and processing stabilisation of the PA6 and PA6 composite with HNT reinforcement. They investigated combinations of primary antioxidant (Irganox 1010) and secondary antioxidants (i.e., Irgafos 168 and TNPP) using specific tests. They reported higher tensile strength, modulus and elongation for PA6/HNT (95/5) composition in the presence of Irganox/TNPP antioxidant than the unstabilised PA6, as well as light colouring of the PA6 and enhanced flow of the composites.

In a study done by Svachova et al. (2017), the effect of HNT on the structure, morphology, mechanical and thermal properties, as well as cell behaviour of the electrospun PCL/Gelatine nanofibres was investigated. The highest improvement of mechanical properties has been achieved in nanofibres with 0.5 wt. % HNT, and the reinforcing effect of HNT was observed up to 3 wt. % HNT. Cytotoxicity using mouse fibroblasts showed that the studied PCL/Gelatine/HNT nanofibres are non-toxic and fully acceptable for medical applications.

Another study, done by Kelnar et al. (2016), deals with the modification of mechanical properties of PCL/PLA system using the MFC concept. This system was modified with HNT, and allowed successful melt drawing because of the modification of rheological parameters of the polymer components. The best results were obtained with 3 phr HNT content. Also, it was observed that the final MFC processing below the melting point of PLA led to an increase in crystallinity of amorphous neat PLA and to an enhanced mechanical performance.

The interest for bio-PA11 has increased considerably, being an environmentally friendly option. In this context, Gorrasi et al. (2018) studied the photo-oxidative degradation processes occurring on PA11 composites based on thymol and HNT, when thymol was entrapped into the lumen of HNT. It's about a synergic effect between them. Thymol avoided the acceleration of oxidative reactions from HNTs, and HNTs worked as nanocontainers determining a slower release of thymol molecules from PA11 bulk. Thus, an improvement of the thermal and mechanical properties during UV irradiation occurred. Hao et al. (2015) focused on the influence of HNT loading on the mechanical and flammability properties of the PA11/FR/HNT nanocomposites. The formula with 25% FR and 2.5% HNT showed the enhancements of mechanical properties, increases of the crystallisation temperature, decreases of PHRR, and UL-94 V-0 performance. Francisco et al. (2019) studied the influence of HNT on the PA11/HNT nanocomposites. They reported the enhancement of the mechanical properties (i.e., tensile strength, flexural strength and modulus), but the decrease of the impact properties.

Table 2 summarise the main properties of HNTs/Polyamide nanocomposites in relation with the principal obtaining routes.

Table 2. HNTs/PA-based nanocomposites

Nano-composites	HNTs modifier	HNT loading (%)	Mixing method	Properties variations (%)(compared to neat resin)	Ref.
HNT/PA11/ SEBS-g-MA (5/85/15)	APTES; SEBS-g-MA	5	Melt blending	HNT/PA11/SEBS-g-MA -Tc: +2.2 -Xc: +24 -Yield stress: -11.2 -elongation at break:-11 -Young's modulus: -8.1 -Notched Charpy Impact strength: +114.5 -storage modulus in flexion: (-45°-+56°C)< sHNT/PA11/SEBS-g-MA -Tc: +2.1 -Xc: +21.7 -Yield stress: -22.8 -elongation at break: +21.5 -Young's modulus:-19.2 -Notched Charpy Impact strength: +683.6 -storage modulus in flexion: (-45°-+56°C)<	Sahnoune et al. 2017a
HNT/PA6/ NBR	(3-glycidyloxypropyl trimethoxysilane)	3, 5, 7	Melt blending	PA6/NBR(60/40)/ModifiedHNT -tensile strength: +(20 ÷ 45)	Paran et al. 2017
HNTs-(disc-shaped) diatomite/PA6	-	0 - 6	Melt blending	PA6/HNT/Diatomite (94/5/1) -tensile strength: +46 -Young's modulus: +23 -elongation at break:-55 -flexural strength: +43.6 -notched impact strength: +45.4	Cheng et al. 2018a

Table 2. (Continued)

Nano-composites	HNTs modifier	HNT loading (%)	Mixing method	Properties variations (%)(compared to neat resin)	Ref.
HNT/PA6	BDP (2, 4%)	5, 10	Melt blending	PA6/HNT/BDP4 -PHRR: -38.9 -TTI: - 8.8 -Tg: -22.3 -tensile strength: -22.1 -elongation at break: -62.8 -Young's modulus: -5.7 PA6/BDP2(in bulk)-loaded HNT/BDP2 (in lumen) -PHRR: -44.1 -TTI: -11.6 -Tg: -11.3 -tensile strength: -13 -elongation at break: -68 Young's modulus: +15.7	Boonkongkaew et al. 2018
HNT/PCL/ Gelatine (PCL:Gelatine = 1 : 1)	-	0.5, 3, 6	Electro-spinning	0.5HNT/PCL/gelatin -pore size(nm): 170-570 -tensile strength: +69.2 -Young's modulus:+63.5 -elongation at break: +249 -Xc: -16.7 3HNT/PCL/gelatin -pore size(nm): 200-700 -tensile strength: +23.1 -Young's modulus:+6.6 -elongation at break: +21.1 -Xc: -22.2	Svachova et al. 2017

Nano-composites	HNTs modifier	HNT loading (%)	Mixing method	Properties variations (%)(compared to neat resin)	Ref.
HNT/PCL/PLA (80/20)	-	0, 3, 5	Melt drawing	PCL/PLA/HNT 80/20/3 DR 1 -tensile strength: -23 -elongation at break: -57.4 -Young's modulus: +156 -tensile impact strength: -60.2 PCL/PLA/HNT 80/20/3 DR 6 -tensile strength: -11.7 -elongation at break: -82.4	Kelnar et al. 2016
HNT/PA11	HNT-Thymol (44.5/55.5 wt%)	5 wt% Thymol+5 wt% HNT; 10 wt% HNT-thymol	Compression molding	PA11+5%Thymol+5%HNT(1G) -Decompos. Max. temp. +5 PA11+ 4.45%HNT+%.55%Thymol (Photo-aging time(day)) (3) -tensile strength: +22.2 -elongation at break: +18.2 (5) - tensile strength: +44.4 -elongation at break: +51.5 (7) - tensile strength: +33.3 -elongation at break: +39.4	Gorrasi et al. 2018
HNT/PA11/FR	FR = additive based on metal phosphinate	2.5, 5, 10	Melt blending	HNT/PA11/FR 2.5/25/32.5 -Tc: +1.4 -Ultimate tensile strength: -25.1 -elongation at break: -94.8 -tensile modulus: +35 -PHRR: -27 -IT: +5.4	Hao et al. 2015

Table 2. (Continued)

Nano-composites	HNTs modifier	HNT loading (%)	Mixing method	Properties variations (%) (compared to neat resin)	Ref.
HNT/PA11		2, 6, 10	Melt blending	2HNT/PA11 -tensile strength: -5.3 -elongation at break: -28.1 -modulus of elasticity: +19.6 -flexural strength: +10.8 -Izod impact resistance: -38.5 6HNT/PA11 -tensile strength: -19.3 -elongation at break: -6.1 -modulus of elasticity: +21.4 -flexural strength: +15.6 -Izod impact resistance: -16.3 10HNT/PA11 -tensile strength: -29.6 -elongation at break: -4.2 -modulus of elasticity: +27.1 -flexural strength: +20.3 -Izod impact resistance: -48.6	Francisco et al. 2019

HNTs as Reinforcement for Polylactic Acid-Based Polymer Composites

PLA is a kind of aliphatic polyester derived from renewable resources. The interest for PLA increased very much due to its good biocompatibility, biodegradability, renewability, recyclability, processability, high optical transparency, and proper mechanical properties. The main applications of PLA are food packaging, automotive parts, and biomedical devices. However, its weaknesses are brittleness (or poor toughness) and low thermal stability. But, the toughness of brittle polymers can be efficiently improved by their blending with soft, ductile polymers with a low Tg (Bucknall and Paul 2009). Many efforts have been made to enhance the properties of PLA using: PCL (a rubbery polymer with low Tg) (Kelnar et al. 2016), ENR (Tham et al. 2016a,b), SEBS-g-MA (Chow et al. 2018, Tham et al. 2016c), fillers like carbon nanotube (Yoon et al. 2009, Villmow et al. 2008), graphene (Valapa et al. 2015), and others. From a decade it is reported that halloysite nanotube can improve the mechanical and thermal properties of polyolefins (Dulebova and others 2018, Jenifer et al. 2018), PAs (Sahnoune et al. 2017a), and of other polymers.

Compatibilising with Grafted Polymer

Tham et al. (2016c) studied the effects of water absorption on the thermal and impact properties of PLA/HNT/impact modifier (SEBS-g-MA, or EBS) at three temperatures. They reported a decrease of diffusion coefficient of PLA, also the activation energy of water absorption of the nanocomposite containing EBS was higher than that containing SEBS-g-MA. The decreases of impact strength, Tg, Tcc and melting temperature of the PLA were observed.

In a study done by Chow et al. (2018) are reported the effects of SEBS-g-MA and EBS on the mechanical and thermal properties of PLA/HNT nanocomposites.

The PLA/HNT6%/EBS5% composite showed higher impact strength compared to SEBS-g-MA counterpart, also Xc showed higher values. EBS was less sensitive to the oxygen atmosphere as compared to SEBS-g-MA.

Pre-Treatment with Coupling Agent

Krishnaiah et al. (2017), studied the effect of silane functionalized HNTs on the morphology, mechanical, dynamic mechanical and thermal properties of HNT-PLA nanocomposites. They reported a significantly increased of tensile strength, tensile modulus, impact strength, storage modulus, tan delta, thermal stability, as well as an enhancement of crystallinity, and Tg of the ASP functionalised HNT reinforced PLA nanocomposites.

Long-term biological applications (i.e., coronary stent) that require high strength for up to 6 month and thereafter biodegradation has been the object of a study done by Chen et al. (2018).

The reinforced PLA/ASP-HNTs composite with 5 wt. % ASP-HNT loading displayed a Young's modulus increased by 25% and retained 74% of this value up to the end of the degradation period. Also, ASP-HNT acted as nucleation agent and increased crystallisation and degradation process of PLA. The degradation in crystalline region of the composite did not occur.

The effects of the addition of amino-modified HNTs on the degradation and properties of mechanically recycled PLA were studied by Beltran et al. (2018). Accelerated ageing and mechanical recycling caused a significant degradation of the neat polymer, a decrease in its intrinsic viscosity, Tcc, thermal stability and microhardness. In contrast, when the amino-modified HNT was used they reported improvements in the above properties due to more effective blocking of the carbonyl groups, as a result of acid-base interactions with the amino groups of the silanised HNTs.

No Treatment

The effect of ENR on the properties of PLA/HNT-based nanocomposites was the object of three studies. Tham et al. 2016a studied the oxygen permeability, thermal properties and light transmittance influenced by ENR and HNTs. They reported a decrease of the oxygen permeability upon 6% HNT, high thermooxidative stability under oxygen atmosphere and maintaining the transparency. Tham et al. 2016b investigated the mechanical properties and water absorption of PLA/HNT2% nanocomposites influenced by the ENR. They reported increases of the impact strength, activation energy of water diffusion, and also the high percentage retention of impact strength after exposure to water absorption. Teo et al. 2016 studied the morphological and thermal properties of PLA/PMMA/HNT nanocomposites. They concluded that incorporation of ENR increased the impact strength, maintained the T_m of the nanocomposite but decreased the X_c.

A study done by Tham et al. (2015) about the water absorption behaviour of PLA/HNT nanocomposites at three temperatures 30, 40 and 50°C revealed that the Ea of water diffusion for PLA increased with the incorporation of HNT and EBS. The T_g, T_m, and T_{cc} for PLA shifted to lower temperatures, and the carbonyl index values increased after water absorption due to the formation of higher amount of carboxylic acid end groups during the hydrolysis process.

The modification of PLA properties by filling with HNT and EBS-treated HNT has been the subject of a study done by Pluta et al. (2017). They founded that the EBS-treated nanofiller nucleated PLA matrix and decreased Tcc, also, the storage modulus was enhanced. In contrast, PLA blended with EBS has showed a decrease in storage modulus compared to quenched PLA. The storage modulus of the EBS-HNT/PLA nanocomposites at 20, and 60°C increased compared to neat PLA, but decreased with temperature. Furthermore, they revealed that the filling of PLA with HNT or EBS-treated HNT has created a barrier for UV-B radiation.

Effect of low volumes of HNTs on mechanical and thermal properties of extruded PLA films has been studied by Chen et al. (2017). They reported that Young's modulus and stiffness of PLA were improved with the introduction of HNT. In contrast, the thermal stability of PLA was adversely affected by HNT due to the presence of voids, which have determined a poor contact between HNT and the PLA matrix.

In a study done by Kaygusuz and Kaynak (2015) were investigated the influences of HNTs on the isothermal and non-isothermal melt crystallisation kinetics of PLA, and crystallinity of injection moulded and annealed specimens. They found that isothermal melt crystallisation indicated a higher crystallinity and lower crystallisation times compared to neat PLA, due to the heterogeneous nucleation effect (HNE) at the addition of HNTs. On the other hand, non-isothermal melt crystallisation indicated that HNE of HNTs determined an increased crystallinity, 10 times higher, and all temperature parameters were shifted to higher values. Furthermore, they revealed that in case of the injection moulded and annealed specimens the highest crystallinity degree of PLA could be obtained with only 1 wt. % HNT.

The uses of HNTs as nano-containers for Lysozyme, a natural protein with antimicrobial activity for packaging with controlled release in food protection, and incorporation into a PLA matrix has been studied by Bugatti et al. (2017). They evaluated the structural characterisation and physical and barrier properties of films obtained. A reinforcement of PLA matrix at low filler loading, and with decreasing of elongation at break at high filler content was reported. Also, they showed that the barrier properties to water vapour were improved.

The production of a cytocompatible, non-toxic and mechanically stable multilayered porous membrane encapsulated with aminoglycoside antibiotic for bone regeneration with a high concentration of drug at the first stage of release and extended drug release has been studied by Pierchala et al. (2018). The membrane consisting of 3 different layers was obtained by electrospinning. They analysed the adsorption capacity of the first porous layer and soaked with an antibiotic, the second layer with HNTs in pores and the third layer comprising HNTs within non-porous PLA nanofibres. They reported drug encapsulation efficiency in the drug loaded complete set of about 53%, and that each fibrous layer exhibited different mechanical properties and loading efficiency.

De Silva et al. (2016) conducted an exploratory analysis of the effects of fabrication methods, solution casting and melt compounding, on the mechanical, crystallinity, morphological, and thermal properties of PLA/HNTs nanocomposites films by varying the HNT content. Also, it was investigated the strain rate dependency of the PLA/HNTs nanocomposites at two rates. They reported a significant influence of the preparation methods on the mechanical properties. The tensile properties increased for both processing methods being higher in case of melt compounding, 5 wt. % HNT being the optimum concentration. Also, melt compounding resulted in a higher crystallinity.

Therias et al. (2017) studied the effect of photooxidative degradation on the PLA-HNT nanocomposites obtained by melt compounding. They reported high tensile strength and Young's modulus, as well as increase in the storage modulus of the nanocomposites. They found that HNT presented a pro-degradant effect on the photooxidation of PLA, due to the presence of chromophore groups and traces of iron.

In a study done by Kim et al. (2016) the PLA/HNT nanocomposites were prepared by melt mixing. The morphology, rheology, as well as the mechanical and thermal properties of the nanocomposites were studied. They showed that the HNTs were well-dispersed in the PLA matrix at lower loading levels and a decrease in tensile strength was observed for higher levels of 5 wt. % HNT contents. Also, they reported that at increased HNT levels the rheological properties of the composites were enhanced, and a link between the shear viscosity as a function of shear rate and the complex viscosity as a function of angular frequency was established using Cox-Merz rule.

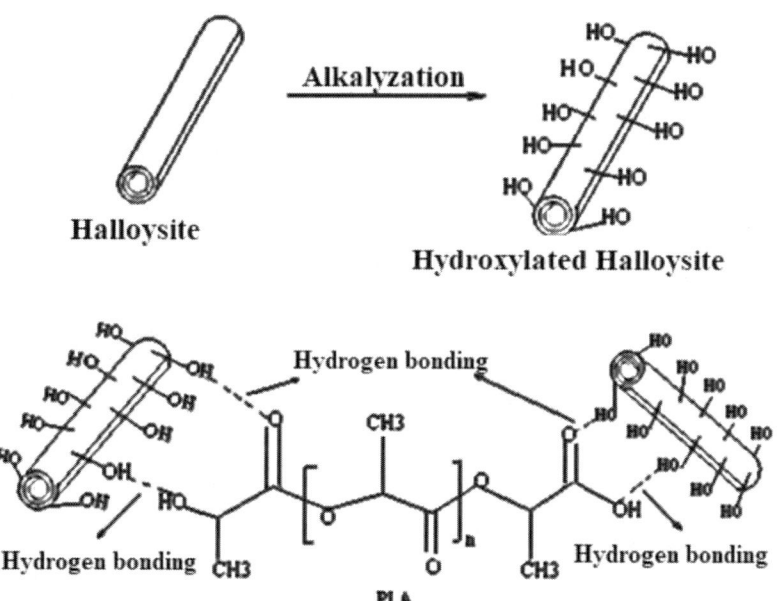

Figure 4. The procedure of HNT alkalysation (adapted from Guo et al. 2016).

Guo et al. (2016) introduced alkalised HNTs in PLA/HNTs nanocomposites (Figure 4) and studied its crystallisation, morphology, mechanical properties, and thermal stability for determining the effect of the surface treatment and use of alkalised HNTs on the characteristics of PLA. They reported that alkalised HNT could act as a nucleating agent for PLA, leading to decreasing T_{cc} and increasing crystallinity. Furthermore, alkalised HNT increased the modulus, tensile strength, and thermal stability of PLA.

An investigation concerning the effect of the HNT concentration as well as the addition of glycerol on the mechanical properties of the PLA nanocomposite films was conducted by Risyon et al. (2016). They reported that films produced without the addition of glycerol exhibited higher tensile strength and elongation at break compared to film with glycerol, due to the fact that glycerol encouraging mobility of the particles in the composite, the intermolecular force was decreased in the bionanocomposite films. Furthermore, glycerol taking part in the interlayer spacing of the clay but not the polymer molecules, poor mechanical properties were obtained.

In a study done by Alakrach et al. (2018), the effect of incorporating of hybrid fillers, HNTs and ZrO_2 nanoparticles into PLA matrix on the mechanical properties was studied. They founded that the addition of ZrO_2 decreased the mechanical properties of PLA/HNTs- ZrO_2.

Table 3 summarise the main properties of HNTs/PLA nanocomposites in relation with the principal obtaining routes.

HNTs as Reinforcement for Polymethylmethacrylate-Based Composites

PMMA is an amorphous, hydrophobic, brittle polymer, having several desirable properties, including good weatherability, high Young's modulus, a low elongation at break, good scratch resistance.

Table 3. HNTs/PLA-based nanocomposites

Nanocomposites	HNTs modifier	HNT loading (%)	Mixing method	Properties variations (%) (compared to neat resin)	Ref.
HNT/PLA	SEBS-g-MA, EBS (5-20 wt%)	6	Melt blending	PLA/6HNT/5SEBS-MA -Ea: +742.3 -Xc: - 7 -Xc (50°C): +33.5 -Charpy impact strength: (30°C) -20.8 (40°C) -12.9 PLA/6HNT/15SEBS-MA -Ea: +696.5 -Charpy impact strength: (30°C) +14 (40°C) - 3 PLA/6HNT/20SEBS-MA -Ea: +713.4 -Xc: -40.8 -Charpy impact strength: (30°C) +7.2 (40°C) -37.3 PLA/6HNT/5EBS -Ea: +992.6 -Xc (50°C): +14.2 -Charpy impact strength: (30°C) -32.3 (40°C) -54.8 PLA/6HNT/10EBS -Ea: +1024.6 -Charpy impact strength:	Tham et al. 2016c

Table 3. (Continued)

Nanocomposites	HNTs modifier	HNT loading (%)	Mixing method	Properties variations (%) (compared to neat resin)	Ref.
				(30°C) +33.6 (40°C) +35.8 PLA/6HNT/15EBS -Ea: +939.7 -Charpy impact strength: (30°C) -32.7 (40°C) -62.1	
HNT/PLA	SEBS-g-MA, EBS (5-20 wt%)	6	Melt blending	PLA/6HNT/5SEBS-MA -tensile strength: -34.8 -elongation at break: -17.4 -tensile modulus: +5.9 -Charpy impact strength: +124.4 (+77% compared to PLA/6HNT) -Tg: -3.7 -Tcc -14.7 (compared to PLA/6HNT) -Xc: -7 -OOT: -7.6 PLA/6HNT/15SEBS-MA -tensile strength: -52.2 -elongation at break: +2.2 -tensile modulus: -29.4 -Charpy impact strength:+82.2 -Tg: +3.4 -Xc: -7 PLA/6HNT/20SEBS-MA	Chow et al. 2018

Nanocomposites	HNTs modifier	HNT loading (%)	Mixing method	Properties variations (%) (compared to neat resin)	Ref.
				-tensile strength: -59.8 -elongation at break: +10.9 -tensile modulus: -35.3 -Charpy impact strength:+84.4 -Tg : +5.1 -Xc: -40.8 -OOT: -7 PLA/6HNT/5EBS -tensile strength: -36 -elongation at break: -23.9 -tensile modulus: - 5.9 -Charpy impact strength: +151.1 (+98% compared to PLA/6HNT) -Tg: -0.7 -Tcc -13 (compared to PLA/6HNT) -Xc: +5 -OOT: +2.1 PLA/6HNT/10EBS -tensile strength: -41 -elongation at break: -28.3 -tensile modulus: 0 -Charpy impact strength:-98.9 -Tg: -1.7 -Xc: -5 PLA/6HNT/15EBS -tensile strength: -51.5 -elongation at break: -21.7 -tensile modulus: 0 -Charpy impact strength: +70	

Table 3. (Continued)

Nanocomposites	HNTs modifier	HNT loading (%)	Mixing method	Properties variations (%) (compared to neat resin)	Ref.
HNT/PLA	APTES	2, 4, 6, 8	Melt blending	-Tg: +0.3 -Xc: -6 4HNT-APTES/PLA -tensile strength: +26.5 -elongation at break: +69.6 -tensile modulus: +13.9 -Impact strength: +39.3 -Tg : +3.1 -Storage modulus: +13.8 -Xc: +30 6HNT-APTES/PLA -tensile strength: +11.9 -elongation at break: + 3.3 -tensile modulus: +14.8 -Impact strength: +41.6 -Tg : +5.8 -Xc: +54.1 8HNT-APTES/PLA -tensile strength: +7.7 -elongation at break: +12 -tensile modulus: +20.7 -Impact strength: +15.9 -Tg (tan δ): +4.9 -Storage modulus: -11.9	Krishnaiah et al. 2017

Nanocomposites	HNTs modifier	HNT loading (%)	Mixing method	Properties variations (%) (compared to neat resin)	Ref.
HNT/PLA	3-aminopropyltriethoxysilane	1, 2, 5	Melt blending	1HNT-APTES/PLA -Tg: -3.5 -Xc : +5.1 -Hemolysis rate : -46.3 2HNT-APTES/PLA -Tg: -3.5 -Xc : +4.5 -Young's modulus : +12,9 -Hemolysis rate : -79.6 5HNT-APTES/PLA -Tg: +3.7 -Xc : +8.7 -Young's modulus : +25 -Hemolysis rate : -62	Chen et al. 2018
HNT/PLA	APTES	2	Melt blending	-	Beltran et al. 2018
HNT/PLA	ENR-50 (epoxidized natural rubber, 50% mol of epoxidation)	2, 6	Melt blending	2HNT/PLA/5ENR -oxygen permeability: -9 -OOT: +19.5 -transmittance (600 nm UV): -60.8 2HNT/PLA/20ENR -oxygen permeability: +32.6 -OOT: -13.4 -transmittance (600 nm UV): -87.4 6HNT/PLA/5ENR -oxygen permeability: -30.3 -OOT: +34	Tham et al. 2016a

Table 3. (Continued)

Nanocomposites	HNTs modifier	HNT loading (%)	Mixing method	Properties variations (%) (compared to neat resin)	Ref.
				-transmittance (600 nm UV): -94.3 6HNT/PLA/20ENR -oxygen permeability: +46.4 -OOT: -17.6 -transmittance (600 nm UV): -97.9	
HNT/PLA	ENR-50 (epoxidized natural rubber, 50% mol of epoxidation)	2	Melt blending	2HNT/PLA/5ENR -tensile strength: -30.2 -elongation at break: - 6.5 -tensile modulus: - 5.9 -flexural strength: - 6.7 -Charpy impact strength:+33.3 - water abs(30°):-15 -: water abs.(40°):-18.3 2HNT/PLA/10ENR -tensile strength: -39.2 -elongation at break: -26.1 -tensile modulus: -17.6 -flexural strength: -36 -Charpy impact strength: +117.8 - water abs(30°): - 2 -: water abs.(40°): +9.2 2HNT/PLA/15ENR -tensile strength: -59.8	Tham et al. 2016b

Nanocomposites	HNTs modifier	HNT loading (%)	Mixing method	Properties variations (%) (compared to neat resin)	Ref.
				-elongation at break: -52.2 -tensile modulus: -17.6 -flexural strength: -63.4 -Charpy impact strength: +384.4 - water abs(30°): -53.9 -: water abs.(40°):-81.9	
HNT/PLA/PMMA (70/30)	ENR-50 (epoxidized natural rubber, 50% mol of epoxidation)	2, 5	Melt blending	5HNT/PLA/PMMA/5ENR -Charpy impact strength: +9.7 -Tg: +4.2 -Xc: -93.7 5HNT/PLA/PMMA/10ENR -Charpy impact strength:+37.2 -Tg: +2.3 -Xc: -88.5 5HNT/PLA/PMMA/15ENR -Charpy impact strength: +113.8 -Tg: +4.4 -Xc: -90.7	Teo and Chow 2016
HNT/PLA	EBS (5-20 wt.%)	2	Melt blending	2HNT/PLA/5EBS -Tg: -2 (40°): -1.5 (50°): -44.3 -Xc: -25.2 (40°): +38.9 (50°): +46.9 -max. Water abs.: (40°): +87.9	Tham et al. 2015

Table 3. (Continued)

Nanocomposites	HNTs modifier	HNT loading (%)	Mixing method	Properties variations (%) (compared to neat resin)	Ref.
				(50°): +32.3 -pH of water (40°): +1.7 (50°): +0.5 -carbonyl index +44 (40°): -3.5 (50°): +27.2 2HNT/PLA/10EBS -max. Water abs.: (40°): +70.3 (50°): +19.4 -pH of water (40°): -0.5 (50°): +1 2HNT/PLA/15EBS -max. Water abs.: (40°): +73.6 (50°): +10.8 -pH of water (40°): +28.9 (50°): 0 2HNT/PLA/20EBS -Tg: -2.4 (40°): -5.9 (50°): -45.3 -Xc: +19.8 (40°): +16	

Nanocomposites	HNTs modifier	HNT loading (%)	Mixing method	Properties variations (%) (compared to neat resin)	Ref.
				(50°): +72.9 -max. Water abs.: (40°): +82.4 (50°): +19.4 -pH of water (40°): +2.7 (50°): -0.5 -carbonyl index +58.6 (40°): +4.4 (50°): +38.4	
HNT/PLA	EBS(0.33, 0.66, 0.75, 1.50)	3, 6	Melt blending	Quenched all : 3HNT/PLA/0.33EBS -Tg: -2.9 -transmittance(600nm):-85.3 3HNT/PLA/0.75EBS -Tg: -1.5 -transmittance(600nm):-60.6 6HNT/PLA/0.66EBS -Tg: -0.2 -transmittance(600nm):-88.8 6HNT/PLA/1.5EBS -Tg: -0.3 -MLM: +0.28 -storage modulus(20°):+20 -storage modulus(60°):+16.7 -loss modulus: (60°):+40 -transmittance(600nm):-95.5	Pluta et al. 2017

Table 3. (Continued)

Nanocomposites	HNTs modifier	HNT loading (%)	Mixing method	Properties variations (%) (compared to neat resin)	Ref.
HNT/PLA	-	2, 5, 10, 15	Melt blending	5HNT/PLA -Tg: -3.4 -Xc: +1.4 -tensile strength at yield: +19.8 -elongation at yield: -23.6 -Young's modulus: + 6.7 -void content: +85.7 10HNT/PLA -Tg: - 3.6 -Xc: -11 -tensile strength at yield: +14.4 -elongation at yield: -2.8 -Young's modulus: +12.1 -void content: +58.8	Chen et al. 2017
HNT/PLA	-	1, 3, 5, 10	Melt blending		Kaygusuz and Kaynak 2015
HNT/PLA	lysozyme	3, 5, 10	Ball milling	3HNT-PLA/Lysozyme -tensile strength: -6.7 -elongation at break: +11.1 -elastic moduli: +23.6 -water vapor: -82.4 5HNT-PLA/Lysozyme -tensile strength: -15.6 -elongation at break: -58.3	Bugatti et al. 2017

Nanocomposites	HNTs modifier	HNT loading (%)	Mixing method	Properties variations (%) (compared to neat resin)	Ref.
				-elastic moduli: +27.1 -water vapor: -87.6 10HNT-PLA/Lysozyme -tensile strength: -17.8 -elongation at break: -65.6 -elastic moduli: +14.3 -water vapour: -92	
HNT/PLA/ gentamicin	(rate of drug release: up to 48 h)		Electro-spinning	Layer 2S = porous electrospun PLA with HNT in pores of the fibres, soaked with Gentamicin sulphate; Complet set 2S = (Layer 1S, Layer 2S, Layer 3S) Complet set 2S/Layer 1S -tensile strength: +105 -elongation at break: +22.2 -Young's modulus: +148.1 -drug encapsulation efficiency: +23.7 Complet set 2S/Layer 2S -tensile strength: +233.5 -elongation at break: +144.4 -Young's modulus: +59.5 -drug encapsulation efficiency: +17.2 Complet set 2S/Layer 3S -tensile strength: +60 -elongation at break: -57.6 -Young's modulus: +458.3 -drug encapsulation efficiency: +110.7	Pierchala et al. 2018

Table 3. (Continued)

Nanocomposites	HNTs modifier	HNT loading (%)	Mixing method	Properties variations (%) (compared to neat resin)	Ref.
HNT/PLA	-	2.5, 5, 7.5 10	Melt blending, Solution casting	2,5HNT/PLA -tensile strength: +7.5 -elongation at break: -8 -Young's modulus: +6.2 5HNT/PLA -tensile strength: +22 -elongation at break: -24 -Young's modulus: +20.8 7.5HNT/PLA -tensile strength: +9.9 -elongation at break: -22.1 -Young's modulus: +16 10HNT/PLA -tensile strength: +7.9 -elongation at break: -31.2 -Young's modulus: +16.7	De Silva et al. 2016
HNT/PLA	-	3, 6, 12	Melt blending	3HNT/PLA -maximum tensile strength: -2.4 -Young's modulus: +8.8 -Xc: -25.6 6HNT/PLA -maximum tensile strength: -2.9 -Young's modulus: +10.1 -storage modulus(60°): +40.6	Therias et al. 2017

Nanocomposites	HNT's modifier	HNT loading (%)	Mixing method	Properties variations (%) (compared to neat resin)	Ref.
HNT/PLA	-	1, 3, 5, 7, 9	Melt blending	-Xc: -7 12HNT/PLA -maximum tensile strength: -3.5 -Young's modulus: +21.5 -storage modulus(60°): +72.8 -Xc : +58.1 1HNT/PLA -tensile strength at yield: -1.2 -storage modulus: +108.9 -loss modulus: + 33 3HNT/PLA -tensile strength at yield: -2 -storage modulus: +211 -loss modulus: +50 5HNT/PLA -tensile strength at yield: -2.9 -storage modulus: +300 -loss modulus: + 83 7HNT/PLA -tensile strength at yield: -7.3 -storage modulus: +455 -loss modulus: +125 9HNT/PLA -tensile strength at yield: -17 -storage modulus: +166.7 -loss modulus: +216	Kim et al. 2016

Table 3. (Continued)

Nanocomposites	HNTs modifier	HNT loading (%)	Mixing method	Properties variations (%) (compared to neat resin)	Ref.
HNT/PLA	NaOH	1, 4, 6.5, 9	Melt blending	1HNT(NaOH)/PLA -$T_{50\%}$: +1.4 4HNT(NaOH)/PLA -Tg: +0.5 -Xc: +32.7 -$T_{50\%}$: +1.7 6.5HNT(NaOH)/PLA -Tg: +1.4 -Xc: +40 -$T_{50\%}$: +2 9HNT(NaOH)/PLA -Tg: +2.5 -Xc: +56.4 -$T_{50\%}$: +2.9	Guo et al. 2016
HNT/PLA/Glycerol (0.5 ml)	-	2, 4, 6, 8	Solution casting	2HNT/PLA -tensile strength: +1.4 2HNT/PLA/Glycerol -tensile strength: +16.3 4HNT/PLA -tensile strength: -7.3 4HNT/PLA/Glycerol -tensile strength: +30 6HNT/PLA -tensile strength: -17.56 6HNT/PLA/Glycerol -tensile strength: -11.4	Risyon et al. 2016

Nanocomposites	HNT's modifier	HNT loading (%)	Mixing method	Properties variations (%) (compared to neat resin)	Ref.
HNT-ZrO$_2$/PLA	HNT-ZrO$_2$ (1, 3wt.%)	2, 4, 6, 8	Solution casting	2HNT/PLA -tensile strength : +23.8 -elongation at break : +16.2 -Young's modulus : +59.8 4 HNT/PLA -tensile strength : +71.5 -elongation at break : +25 -Young's modulus : +202.3 6 HNT/PLA -tensile strength : +57.1 -elongation at break : +11.8 -Young's modulus : +175.4 8 HNT/PLA -tensile strength : +33.3 -elongation at break : + 2 -Young's modulus : +73.4 HNT-1ZrO$_2$/PLA -tensile strength : -22.2 -elongation at break : -16.3 -Young's modulus : -25.5 HNT-3ZrO$_2$/PLA -tensile strength : -33.3 -elongation at break : -65.1 -Young's modulus : -31.8	Alakrach et al. 2018

The nanocomposites based on PMMA offer the potential for reduced gas permeability and increased heat resistance. They have diverse applications in biomedical as drug carrier (Bettencourt and Almeida 2012), dentistry (Abdallah 2016), in automotive industry (Vuluga et al. 2018), optical, solar, sensor, nanotechnology (Ali et al. 2015).

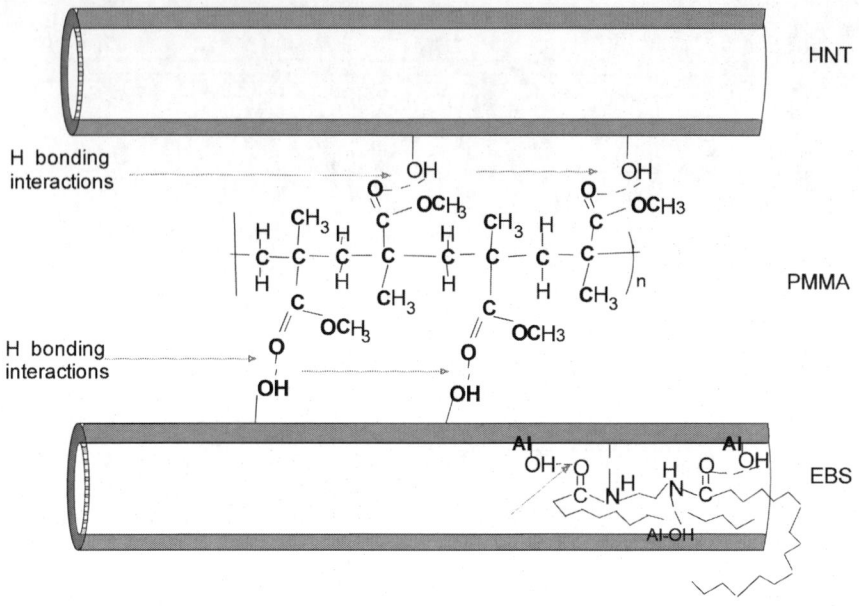

Figure 5. Schematic structures of the interactions between PMMA/HNT-EBS in the composite.

For obtaining PMMA/HNT-EBS nanocomposites for the automotive polymer parts industry, simple polymer processing routes (mixing, extrusion and injection moulding) were used by Vuluga et al. (2018) (Figure 5).

The objective was to improve nanomechanical properties (i.e., scratch resistance, balance between tensile strength and impact resistance) without decreasing other ones. They investigated the relationship between morphological/structural and tribological properties of PMMA nanocomposites. They reported a synergistic effect between HNT and EBS in the PMMA matrix, attained by the phase distribution resulted from the selective interaction between partners and favourable processing conditions.

The initial processing route temperature of HNT with EBS can tune the masterbatche structure and final PMMA nanocomposites properties. The key role was the smart choice of the temperature used for influencing the EBS crystallizing in alpha form, favourable for the later interaction with PMMA. Also, the interaction between EBS and HNT (then EBS-PMMA) was strong enough during melt processing, and high shear process promoted a debonding and some of the HNTs were found in smaller dimensions.

Pal (2016a) studied the thermo-mechanical and rheological performances of HNTs reinforced polymer nanocomposites in comparison with carbon nanotubes (CNTs) and carbon nanofibres (CNFs) reinforced polymer nanocomposites properties. They revealed that dispersion of HNTs and CNFs has been found superior than CNTs. The tensile properties of composites with HNTs are very close to those of CNT-PMMA composites. From DMA, they showed that the compatibility of different nanofillers depends on the variations of frequency and strain on the structure.

Abdallah (2016) studied the effect of incorporating HNTs into PMMA on its flexural strength, hardness and Young's modulus. He concluded that addition of lower concentration of HNTs (0.3 wt. %) to denture base materials could improve some of their mechanical properties.

Table 4 summarise the main properties of HNTs/PMMA nanocomposites in relation with the principal obtaining routes.

HNTs as Reinforcement for Polyolefin-Based Polymer Composites

Fillers are efficient ways to reinforce polymers, improving other properties like dimensional stability and scratch resistance. In organic/inorganic composites, both the dispersion of the fillers and the interface between the filler and the polymer are extremely important for the properties of composites. Even more important is the compatibility between polyolefins and inorganic fillers, due to the high difference in polarity and the fact that polyolefins are chemically inert. This results in poor dispersion level and in a poor interfacial bonding between the filler and the polymer matrix.

Table 4. PMMA-based nanocomposites

Nano-composites	HNTs modifier	HNT loading (%)	Mixing method	Properties variations (%) (compared to neat resin)	Ref.
HNT/PMMA	EBS (N,N'-ethylenebis (stearamide))	2	Melt blending	HNT/PMMA -tensile strength: -8.3 -axial strain: 0 -impact strength: -1.3 Young's modulus: -10.6 -reduced modulus: -3.4 -contact depth: -2.1 -scratch penetration depth: -18.7 HNT-EBS (80)/PMMA tensile strength: -1.7 -axial strain: +4.5 -impact strength: +17.2 Young's modulus: -11.5 -reduced modulus: -0.8 -contact depth: +3.2 -scratch penetration depth:-4.3 HNT-EBS (120)/PMMA tensile strength: -10 -axial strain: +36.4 -impact strength: 0 Young's modulus: -19.2 -reduced modulus: -5.4 -contact depth: -8.5 -scratch penetration depth: -24.9 HNT-EBS (160)/PMMA	Vuluga et al. 2018

Nano-composites	HNTs modifier	HNT loading (%)	Mixing method	Properties variations (%) (compared to neat resin)	Ref.
				tensile strength: -5 axial strain: +13.6 -impact strength: +13.4 Young's modulus: -17.4 -reduced modulus: +0.6 -contact depth: -2.1 -scratch penetration depth: -34.2	
HNT/PMMA	-	3	Melt blending	-tensile strength: +71.6 -elongation at break: + 5.7 -modulus: +16.7	Pal 2016a
HNT/PMMA	-	0.3, 0.6, 0.9	Casting in stone moulds	HNT/PMMA0.3 -Young's modulus: +12.6 -flexural strength: + 7.6 -surface microhardness:+11.7 HNT/PMMA0.6 -Young's modulus: -12.8 -flexural strength: -22.9 -surface microhardness: -7.7 HNT/PMMA0.9 -Young's modulus: -13.5 -flexural strength: -25.7 -surface microhardness: - 8.2	Abdallah 2016

Table 4. (Continued)

Nano-composites	HNTs modifier	HNT loading (%)	Mixing method	Properties variations (%) (compared to neat resin)	Ref.
HNT/ PLA/PMMA (70/30)	ENR-50 (epoxidized natural rubber, 50% mol of epoxidation)	2,5	Melt blending	5HNT/PLA/PMMA/5ENR -Charpy impact strength: +9.7 -Tg: +4.2 -Xc: -93.7 5HNT/PLA/PMMA/10ENR -Charpy impact strength:+37.2 -Tg: +2.3 -Xc: -88.5 5HNT/PLA/PMMA/15ENR -Charpy impact strength: +113.8 -Tg: +4.4 -Xc: -90.7	Teo and Chow 2016

To overcome these problems modification of both fillers and polymers is necessary. So, many alternative methods were used, i.e., pre-treatment of fillers with coupling agents as well as compatibilizing the polymer with a suitable grafted polymer (Qiao et al. 2017), organically modifying of fillers (Mousavi et al. 2019), or modifying of fillers with branched polyethyleneimine (Kubade and Tambe 2016).

Compatibilising with or without Grafted Polymer

Singh et al. (2016) incorporated HNTs into HDPE by melt mixing and investigated the effect of HNTs and compatibiliser loading on morphology and rheological properties of nanocomposites. In case of HDPE/HDPE-g-MA/HNT (80/10/10) they reported the increases of drawability, melt strength, tensile strength, tensile modulus, and flexural strength, but also the decrease of elongation at break, and impact strength. The incorporation of HDPE-g-MA has improved both the dispersion of HNTs and the interfacial adhesion between matrix and filler.

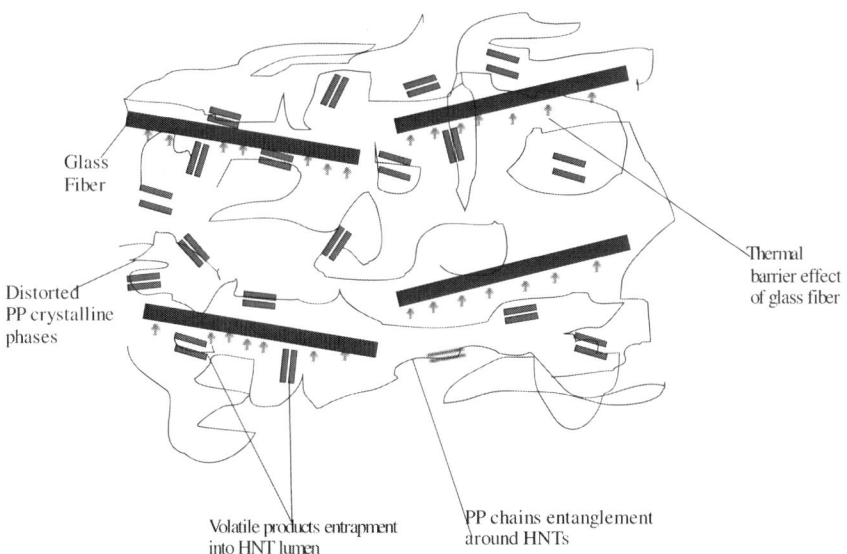

Figure 6. Schematic of barrier and entrapment effect of the fillers for flame resistance of the composite (adapted from Jenifer et al. 2018).

Dulebova and others (2018) investigated the influence of HNTs and PE-g-MA within LDPE matrix on the mechanical properties. They reported an increase of tensile strength, Young's modulus, strain at strength, a decrease of Charpy impact strength (9.5%) and no significant effect on hardness.

The mechanical, thermal, crystallisation, and flame retardant behaviour of multiscale fillers in a PP matrix has been studied by Jenifer et al. (2018). They found that the synergistic effect of micro (GF) and nano (HNTs) fillers in PP, along with presence of compatibiliser has improved both the mechanical and flame retardant properties of the matrix. This is due to the combined effects of barrier and entrapment exerted by the fillers against the thermal transport of volatile products (Figure 6).

Dielectric properties of nanocomposites based on PP, PLA and HNT were investigated by Rajan et al. (2017), by means of mechanical and dynamical mechanical analysis. They found that 6 wt. % of HNT is the optimum loading of nanofiller to the base matrx. Kubade and Tambe (2017) introduced PEI-modified HNTs into PP/ABS blends in presence of dual compatibiliser. An important improvement in tensile modulus, impact strength and thermal stability of PEI-modified-HNTs –filled 80/20 PP/ABS blends has resulted due to the refinement in matrix droplet morphology, increase of crystallinity of PP phase, formation of β form of PP crystals, selective localisation and improved dispersion of PEI-modified HNTs.

Pre-Treatment with Coupling Agent

In a study done by Qiao et al. (2017), HNT/UHMWPE nanocomposite films with Vaseline as swelling agent were prepared and HNTs were surface modified with different coupling agents of oleic acid and KH550 by ultrasonication. They studied the morphology, microstructure, thermal properties, mechanical properties and wettability of these nanocomposites films and concluded the well dispersed nanotubes remarkably promote the crystallite growth, improve the thermal stability and wettability, as well as reinforce the mechanical strength of the nanocomposites.

Liu et al. (2017) introduced pre-treated HNTs, with coupling agents KH550 or EPB, in PP. From mechanical and tribological characterisations they concluded that the tensile, flexural strength and notched impact strength of the nanocomposites were somewhat improved, but the wear resistance was obviously enhanced.

No Treatment

In a study done by Cheng et al. (2018a), the HNTs/disc-shaped diatomite mixture was used to study the synergistic reinforcing effect of the filler in PP matrix. The mechanical performance of the composites was investigated as well as the structure of HNTs/diatomite mixture filler by XRD and SEM. The results showed that the optimally reinforcing performance appears at the HNTs/diatomite mixture proportion of 3:3, suggesting that the above mixture filler could exert a synergistic effect on reinforcing polymer.

HNTs and nanoplate Kaolinite were introduced into PP containing IFR by Sun et al. (2018). The flammability of the PP composites was characterized by LOI, vertical burning (UL-94) and cone calorimeter tests. The results showed that the composite with 75 wt. % PP, 23.5 wt. % IFR, and 1.5 wt. % (Kaolinite/HNT = 9/1) has an increased LOI. The combustion process was prolonged and the HRR was also reduced sharply. The fire resistance and thermal stability were synchronously improved.

Michlik et al. (2016) studied the spinning, structure and mechanical properties of nanocomposite PP/HNT fibres in a discontinuous technological process. They found that the addition of HNT in PP fibres prevents creating of regularity and the arrangement of the supermolecular structure of the oriented nanocomposite fibres reduces crystallinity. This lead to the decrease in mechanical properties of drawn PP/HNT fibres.

Food packaging materials with barrier properties were studied by Tas et al. (2017), based on HNTs and LDPE. They reported that HNTs/PE films with 1 wt. % HNTs content have presented the best oxygen and water vapour barrier properties, demonstrated by the improved shelf life of fruits and chicken samples packaged with these.

Table 5. HNTs/Polyolefin-based nanocomposites

Nanocomposites	HNTs modifier	HNT loading (%)	Mixing method	Properties variations (%) (compared to neat resin)	Ref.
HNTs/HDPE	HDPE-g-MA (5,10,15%)	1, 3, 5, 10	Melt blending	HNTs/HDPE-g-MA/HDPE (10/10/80) -drawability: +15 -melt strength: +44 -tensile strength: +6 -tensile modulus: +9 -flexural strength: +16 -elongation at break: -10 -impact strength: -49	Singh et al. 2016
HNTs/LDPE	PE-g-MA (5 wt %)	2, 4, 6	Melt blending	2HNTs/LDPE/PE-g-MA -tensile strength: +7.9 -strain at strength: -6.3 -Young's modulus: +0.6 -impact strength: -6.8 4HNTs/LDPE/PE-g-MA -tensile strength: +7.6 -strain at strength: -9.4 -Young's modulus: +3 -impact strength: -9 6HNTs/LDPE/PE-g-MA -tensile strength: +9 -strain at strength: -15.2 -Young's modulus: +1.8 -impact strength: -9.5	Dulebova et al. 2018
HNT/PP/FS (74.5/20)	PP-g-MA	2.5	Melt blending	PP/HNT/PP-g-MA (94.5/2.5/3) -tensile strength: +15.4 -tensile modulus: +9.7 -elongation at break: -9.1	Jenifer et al. 2018

Nanocomposites	HNTs modifier	HNT loading (%)	Mixing method	Properties variations (%) (compared to neat resin)	Ref.
HNT/PP/PLA/ PP-g-MA (0-10/80/20/3)	-	0 - 10	Melt blending	-Xc : +13.7 PP/GF/HNT/PP-g-MA (74.5/HNT/20/2.5/3) -tensile strength: + 85.8 -tensile modulus: +144 -elongation at break:-66.7 -Xc : +18.9 -Tg +77 -LOI +28.6	Rajan et al. 2017
HNT/PP/ABS (3/69.6/17.4)	Branched PEI PP-g-MA(5 wt.%), SEBS-g-MA (5 wt.%)	3	Melt blending	PP/PLA/PP-g-MA/6HNT -tensile strength : +10.2 -tensile modulus: +6.8 -elongation at break: 0 -storage modulus at 21°C: +8 -Tg of PLA : +3.4 -crystallinity: +8.3 -Tg : +1.1 -tensile strength: +26.6 -tensile modulus: +28.8 -impact strength: +38.5.	Kubade and Tambe 2017
HNT/UHMWPE (UHMWPE matrix : UHMWPE/Vaseline = 10/90 wt%/wt%)	Oleic acid, KH550 coupling agents (2 wt% to the HNT mass)	1, 2, 4, 8	Melt blending	2%HNT/UHMWPE -tensile strength Longitud Dir.: +27 Transvers Dir.: +31 -Contact angle: -2.5	Qiao et al. 2017
HNT/PP	KH550, or EPB	5	Melt blending	PP/HNT/KH550 (100/5/3) -Mass loss(TG) : -1.87 -Contact angle: +100 -tensile strength: +9.32 -flexural strength: +1.09	Liu et al. 2017

Table 5. (Continued)

Nanocomposites	HNTs modifier	HNT loading (%)	Mixing method	Properties variations (%) (compared to neat resin)	Ref.
				-notched impact strength : +35.64 -wear rate: -34.7 PP/HNT/EPB (100/5/3) -Mass loss(TG) : -2.48 -Contact angle: +67 -tensile strength: +13.37 -flexural strength : +7.15 -notched impact strength : +22.55 -wear rate: -27.8	
HNTs-(disc-shaped) diatomite/PP	-	0 - 6	Melt blending	PP/HNT/Diatomite (94/3/3) -tensile strength: +16 -Young's modulus: +38 -elongation at break: -30 -flexural strength: +46 -notched impact strength: +42.7	Cheng et al. 2018a
HNT/PP/IFR (1.5/75/23.5)	Kaolinit (1.5-0.15 wt %)	0.15 - 1.5	Melt blending	PP/IFR/(9Kaol:1HNT) -LOI : +105 -PHRR : -460 -tensile strength : -31.7 -elongation at break: -19	Sun et al. 2018
HNT/PP/IFR (1.5/75/23.5)	Kaolinit (1.5-0.15 wt %)	0.15 - 1.5	Melt blending	PP/IFR/(9Kaol:1HNT) -LOI : +105 -PHRR : -460	Sun et al. 2018

Nanocomposites	HNTs modifier	HNT loading (%)	Mixing method	Properties variations (%) (compared to neat resin)	Ref.
HNT/iPP	-	1 - 4	Melt blending, drawing	-tensile strength : -31.7 -elongation at break: -19 HNT/iPP undrawn -elongation: +50 -tenacity: -45 HNT/iPP drawn -Young's modulus: -35 -elongation: -35	Sun et al. 2018 Michlik et al. 2016
HNT/LDPE	-	1, 3, 5	Melt blending	HNT/LDPE 1%HNT -crystallinity: +20 -OTR: -22 -WVTR: -32 -weight loss of strawberries: -53 -tensile strength: +2.2 -elongation at break: +3.7 HNT/LDPE 5%HNT -Xc : +0.7 -ethylene absorption capacity, at 1 mbar: +20 -weight loss of strawberries: -31 -tensile strength: -17 -elongation at break: -4.3	Tas et al. 2017
HNT/LLDPE/stabilisers+antiox (1-7/94-88/5)	-	1, 3, 7	Melt blending	7%HNT/LLDPE -tensile strength: +9.6 -tensile modulus: +21 3%HNT/LLDPE -tensile strength: +5.4	Cermak et al. 2016

Cermak et al. (2016) studied structural and mechanical influences of HNTs in the LLDPE-based nanocomposites used in the cable industry. They found that up to 3 wt. % non-treated HNTs content presented a sufficient dispersion, despite a certain interphase tension between the non-polar base polymer and the highly polar HNTs.

Table 5 summarise the main properties of HNTs/Polyolefin nanocomposites in relation with the principal obtaining routes.

HNTs as Reinforcement for Polystyrene-Based Composites

Polystyrene, a synthetic aromatic hydrocarbon polymer, can be available as a solid plastic, or in the form of a rigid foam material. The solid plastic form of PS can be general-purpose PS (GPPS), and HIPS type. GPPS, also known as crystal-clear PS, is fully transparent, rigid, fairly brittle, amorphous, nonpolar and is widely used in food packaging, and medical device applications. However, GPPS has several limitations. It has poor oxygen and UV resistance, is rather brittle, below its Tg, it has medium to high tensile strength and low impact strength. Some of its drawbacks can be overcome by copolymerisation with other monomers (polymer database 2019). HIPS is impact resistant, has a matte finish and is a preferred material for thermoforming. Among the different types of foam are expanded polystyrene (EPS) and extruded polystyrene (XPS). The EPS is thermally insulating and has high impact resistance.

In a study done by Mousavi et al. (2019), an aspen-wood-derived biocrude (WB) compatibiliser and dispersing agent for HNT in a HIPS polymeric matrix was used. They reported the mechanism through which WB compound, attached to the HNT surface, interact with HIPS polymer. Using thermogravimetric and rheological analyses they showed that the aromatic compounds of WB benefit from the possibility of covalent interactions with polymer during processing. They stated that the stability of the compounds formed by this way can explain the enhanced compatibility of WB-treated HNT within the matrix of HIPS.

Table 6. PS-based nanocomposites

Nano-composites	HNTs modifier	HNT loading (%)	Mixing method	Properties variations (%)(compared to neat resin)	Ref.
HNT/HiPS	Wood-derived bio-crude (2-hydroxy-4-methylbenzoic acid, or cyclohexanone, 2-ethyl-)	5	Melt blending		Mousavi et al. 2019
HNT/PS/PA11 (80/20; 60/40)	P(S-co-MAPC$_1$(OH)$_2$)= styrene/(methacryloyloxy)methyl phosphonic acid copolymer	3, 10	Melt blending	HNT/PS/PA11 (10/60/40) -Tc: +3 -Xc: +2.6 HNT/PS/PA11 (10/80/20) -Tc: +1.9 -Xc: -4	Sahnoune et al. 2017b
HNT/PS	-	1, 3, 5, 10	Solution casting	1HNT/PS -tensile strength: +11.4 -elongation: -8.7 -Young's modulus: +12 3HNT/PS -tensile strength: +28.6 -elongation: -8.7 -Young's modulus: +29.2 5HNT/PS -tensile strength: +54.3 -elongation: -13 -Young's modulus: +55 10HNT/PS -tensile strength: +48.6 -elongation: -17.4 -Young's modulus: +49.3	Buruga et al. 2018

Table 6. (Continued)

Nano-composites	HNTs modifier	HNT loading (%)	Mixing method	Properties variations (%)(compared to neat resin)	Ref.
				Pore volum -Microporous membrane: 0.229 cc/g -Nanoporous membrane: 0.081 cc/g	
HNT/PS	SMA	5, 10, 15	Melt blending	5HNT/PS/SMA -tensile strength: -2.2÷7.3 10HNT/PS/SMA -tensile strength:-7.6÷14.9 15HNT/PS/SMA -tensile strength: -20	Arat and Uyanik 2017

Other investigations (Sahnoune et al. 2017b) studied the influence of HNT modification, via grafting of P(S-co-MAPC$_1$(OH)$_2$) copolymer chains at the inner surface, on the morphology of PS/PA11 blends (80/20 and 60/40 wt. %). A change in HNT location after his modification appearing, because HNT migrate from the core of the PA11 phase to the matrix polymer blend interface. Also, thermal stability of the polymer matrix is enhanced after HNT modification mainly in 60/40 blend, and HNT act as a nucleating agent.

Buruga et al. (2018) studied the use of PS as a membrane matrix, through the deposition of HNTs into their pores, and investigated their performance in wastewater treatment in relation to effects of the solvent, clay loading, and sonication time. They reported that membranes containing HNTs provided improvments in many directions (morphological, mechanical, and thermal properties along with good hydrophilicity, antifouling properties, good water flux, and rejection/recovery rates).

In a study done by Arat and Uyanik (2017) was investigated the effect of surface modification (with SMA) on the distribution of HNTs and on the physical properties of a polymer matrix. They revealed that the nanotubes were homogeneously dispersed in the polymer matrix due to surface modification, and the modified HNTs were improved the thermal stability of the nanocomposites. In contrast, the tensile strength of the nanocomposites was decreased.

Table 6 summarise the main properties of HNTs/PS nanocomposites in relation with the principal obtaining routes.

HNT as Reinforcement for Various (PC/EVA/PEG/PTFE) Polymer-Based Composites

PC is an engineering thermoplastic having very good mechanical properties, high stiffness and modulus, good dimensional stability, and very good thermal properties. The drawbacks of PC include the susceptibility to some organic solvents and poor resistance to abrasion (Wu et al. 2007). Its main applications are in the field of automotive industry, medical, etc. PEG is a biocompatible, non-toxic polymer with good water solubility, and chemical inertness.

EVA is the copolymer of ethylene and vinyl acetate (VA). The weight percent VA may vary from 10 to 40%. EVA copolymer has good gloss, low temperature toughness, stress crack and UV resistance. It is used in biomedical engineering applications as a drug delivery device, in coatings formulations, flexible packaging, membranes, hose and tube, and footwear. Its drawbacks are low tensile strength, thermal stability, and high flammability. To overcome these drawbacks suitable nanofiller must be incorporated in EVA (Bidsorkhi et al. 2015).

PTFE is a synthetic fluoropolymer that has many applications. PTFE is hydrophobic, has low coefficient of friction, low outgassing, high temperature capability, and high chemical inertness.

Compatibilising with Grafted Polymer

A study done by Pal et al. (2016b) explores the synergistic effect of HNTs along with MA-g-PE on the physical, mechanical and thermo-mechanical properties of PC/COC polymer blend system. They reported that polarity appears in nonpolar COC polymer after functionalisation with polar MA-g-PE, and this polarisation helps the natural clays to form secondary bonds with both of the blend polymers and dispersed in polymer matrix.

Pre-Treatment with Coupling Agent

In a study done by Cheng et al. (2018b) pristine HNTs were functionalised with three different modifiers and the mechanical and tribological properties of PTFE nanocomposites filled with modified HNTs were investigated. They reported improved mechanical properties of the modified HNT/PTFE nanocomposites, and this improved the workability of PTFE. Also, they noted remarkable increase in wear resistance relative to pure PTFE.

No Treatment

Mehdi et al. (2018) proposed a novel drug delivery system. By 6-arm PEG-NH2 grafting on the external surface of HNTs, followed by conjugation of biotin as a vector for the selective delivery of cytotoxic compounds to the cancer cells.

They obtained HNTs-g-PEG-CDs-Biotin that exhibited a superior loading capacity regarding quercetin anticancer agent, with a low toxicity and high biocompatibility.

Bidsorkhi et al. (2015) studied the preparation of EVA/HNT nanocomposites and comprehensive evaluation of HNT content on mechanical, thermal, water uptake, as well as oxygen barrier properties of the nanocomposites. They reported an improving of the thermal stability by the incorporation of up to 3 wt. % HNT, as well as of ductility and toughness. In contrast, the water uptake and oxygen permeation were decreased.

Gamini et al. (2017) studied the thermal and mechanical performances of PTFE nanocomposites containing HNTs as functional fillers. They reported increases of the degree of crystallinity, storage modulus, loss modulus, tan delta, yield tensile strength, tensile modulus, and impact strength.

Table 7 highlights the relation between obtaining routs and the properties profile for the main HNT applications.

Halloysite Nanotubes in Delivery Systems for Biomedicine and Pharmacotherapy

HNTs in Cancer Therapy

Cancer topic, can be easily distinguished among other ones, by the high rank of influence on citation, on other publications, communications or even patents. In the last decades "nano" and molecular vehicles were developed either for treatment attempts or for imaging purposes. In both cases HNT played a role and still has unexplored pathways for future applications. The latest trends highlighted the possibility of generating multifunctional HNT nanoparticles with both diagnostic and therapeutic role.

Table 7. PC/EVA/PEG/PTFE –based nanocomposites

Nanocomposites	HNTs modifier	HNT loading (%)	Mixing method	Properties variations (%)(compared to neat resin)	Ref.
HNT/PC/cyclic olefin copolymer(PE-g-MA) (70/30)	-	1	Melt blending	1HNT/PC/COC-(PE-g-MA) -Xc: +10.1 -weight loss(5%): +1.1	Pal et al. 2016b
HNT/PTFE	MPS SDS APTES		Compression moulding and sintering	HNT-SDS -tensile stress: 0 -elongation at break: -56.6 -Young's modulus:+226 -Wear rate: -96 HNT-PMMA -tensile stress: -11.4 -elongation at break: -74.1 -Young's modulus:+274 -Wear rate: -96 HNT-COOH -tensile stress: +13.9 -elongation at break: -54.2 -Young's modulus:+263 -Wear rate: -99	Cheng et al. 2018b
HNT-g-PEG-carbon dots-Biotin (Vitamin H)	PEG-NH$_2$, carbon quantum dots, Biotin	-	Solution blending		Mehdi et al. 2018
HNT/EVA	-	1, 3, 5	Solution casting	1HNT/EVA -Xc: -8.7 -tensile strength: +11 -elongation at break:+61.3 -Young's modulus: +32.8 -water absorbency(24h): -53.8	Bidsorkhi et al. 2015

Nanocomposites	HNTs modifier	HNT loading (%)	Mixing method	Properties variations (%)(compared to neat resin)	Ref.
				-oxygen permeability:-9.4 3HNT/EVA -Xc: -10.6 -tensile strength: +40.6 -elongation at break:+62.4 -Young's modulus: +48 -water absorbency(24h): -74.2 -oxygen permeability: -23.1 5HNT/EVA -Xc: -15.4 -tensile strength: +21.5 -elongation at break: +26.7 -Young's modulus: +81.8 -water absorbency(24h): -65.6 -oxygen permeability:-4.6	
HNT/PTFE	-	2 - 10	Compression moulding and sintering	2HNT/PTFE -yield tensile strength: +18.5 Young's modulus: +69.2 -impact strength: +8.3 -Tg: -0.8 -Xc: +10.4 4HNT/PTFE -yield tensile strength: +18.3 Young's modulus: +112.2 -impact strength: +22.8	Gamini et al. 2017

Table 7. (Continued)

Nanocomposites	HNTs modifier	HNT loading (%)	Mixing method	Properties variations (%)(compared to neat resin)	Ref.
HNT/PTFE	-	2 - 10	Compression moulding and sintering	-Tg: -0.05 -Xc: +27.1 6HNT/PTFE -yield tensile strength: +20.7 Young's modulus: +94.3 -impact strength: + 3.7 -Tg: -0.8 -Xc: +12.5 8HNT/PTFE -yield tensile strength: +17.3 Young's modulus: +130.6 -impact strength: +8.4 -Tg: -1 -Xc: +24.99 10HNT/PTFE -yield tensile strength: +30.6 Young's modulus: +147.5 -impact strength: +18.3 -Tg: -0.98 -Xc: +29.2	Gamini et al. 2017

Grims et al. (2018) reported a HNT double chemical modification with folic acid (FA) and fluorescein isothiocyanate (FITC). Such compounds were proven to possess selectivity and high contrast for imaging purposes (in vitro trials on CT-26 colon cancer cell line). Another report (Khodzhaeva et al. 2017) highlighted elevated antitumoral activity for modified HNT using ribonucleases (binase) from *Bacillus pumilus*. Moreover, HNT had a catalytic effect for binase activity. Binase-HNT was reported with a doubled activity in comparison with binase alone against human colon cancer cells (Khodzhaeva et al. 2017).

Other approaches followed the in vitro activity using polycations for HNT modification (Tarasova et al. 2019). Different polycations like poly(ethylenimine), poly(diallyldimethyl-ammonium) and poly (allylamine) can be used for modification of HNT. The procedure can be done in aqueous colloidal state by 2:1 vol ratio (20 mg mL−1 HNT, 10 mg/mL polyelectrolytes) (Tarasova et al. 2019). Polycations are able to stabilise HNT particles in the colloidal state. Such structures however had negative effects on cytotoxicity and supplementary investigations are necessary to evaluate the biological mechanism (Tarasova et al. 2019).

Yang et al. 2016 proposed a robust approach, by using chitosan oligosaccharide for HNT modification. This modification allowed the loading of doxorubicin. The final particles have been shown to be effective in 4T1- bearing mice, with low cytotoxicity (Yang et al. 2016). Recent advances (in vitro and in vivo) improved the doxorubicin loading (in HNT up to 22 %), antitumor efficacy and biocompatibility (Li et al. 2018)

A hybrid approach was reported by Dzamukova et al. (2015) with high activity against human lung carcinoma cells (A549) in comparison with hepatoma cells (Hep3b). The innovation consists in the blocking of HNT lumen, with dextrin after loading with brilliant green (Dzamukova et al. 2015). The dextrin caps can open during the particles uptake next to tumour cells. This approach highlighted also a good selectivity and a low toxicity of the HNT against healthy cells.

The multilayer modification of HNT (i.e., PEGylated and folate derivatised HNT) allow doxorubicin loading and a pronounced biocompatibility (Wu et al. 2018). This approach improves the release of doxorubicin from the nano-carrier (slow release approx. 35h), improves the antitumoral activity (breast cancer) and a better bio-availability in the entire body, due to folate derivatisation.

Grafting chitosan and curcumin loading was proven as an interesting bio-based approach with high biocompatible products (Liu et al. 2016). The report claims a stunning over 90% entrapping ability of curcumin the modified HNT. In this case is one of the few report with clear proves and indications about the net loading capacity (only 3.8% inside or in the lumen *loci*; the rest being adsorbed on the surface). The product showed high activity on different cancer cells and high hemocompatibility and stability (Liu et al. 2016).

Another study focused on the mechanism involved by the intracellular pathway with HNT carries (Liu et al. 2019). In A549 living cells, labelled HNT was introduced by both clathrin- and caveolae-dependent endocytosis, indicating an actin- and microtubule-associated process at the Golgi apparatus and lysosome level. These results helped for drawing the mechanism and explained why HNT are so effective in improving the antitumoral compounds activity. HNT with gemcitabine was used to verify the overall kinetic (Liu et al. 2019).

Sodium polystyrene sulphonate, modified HNT was used to capture MCF-7 cells reaching 92% within 3 h (from artificial blood samples) (Zhao et al. 2017). This approach was proven very promising for cancer patients' monitoring.

When multifunctional HNT carriers are needed (Hu et al. 2017) in vitro studies underlined doxorubin release even after 79 hours, with benefits on HNT endocytosis and with an overall acceleration of the cancer cell apoptosis. Other polymer HNT modifications combined with magnetic functionalisation of HNT were able to release norfloxacin over 60 hours (Fizir et al. 2017). In the same manner gold functionalised HNT hybrid particles provided new platforms for chemo-photothermal cancer therapy (Zhang et al. 2019). These later results explain a wider action of HNT since even neat particles can be able to "capture" cancer cells (MCF-7) from ethanol dispersion 92.5 wt. %) with efficiencies over 90% (He et al. 2018). Other report proved the way of action by using polystyrene sulphonate on

HNT surface; indicating a mechanism useful also for biosensors applications in cancer therapy (Liu et al. 2016).

HNTs in Pharmaceutical

Several studies were conducted to evaluate the HNT ability for general pharmaceutical applications. A HNT-functionalised cyclodextrin was used as a dual drug delivery system for load and release of silibin and curcumin (Massaro et al. 2016a). As excipient, HNT was proven very useful for tablet processing and for efficient load of nifedipine (Yendluri et al. 2017). Covalent functionalisation of HNT can be driven (Yendluri et al. 2017) as dual-responsive nanocarriers for curcumin. This approach allows the controlled release depending on intracellular glutathione and pH conditions. This example explores the possibility of chemical grafting for stimuli-responsive applications (Massaro et al. 2016b).

Antifungal hybrid nanocarriers were obtained by attaching allyloxy β-cyclodextrin (Massaro et al. 2018b). The obtained profile could be exploited in oral Candidiasis and future hydrogel systems for topic administration.

HNT was used also for insulin loading (Massaro et al. 2018b). The estimated loading capacity was around 2.6 wt. % and the in vitro release was over 120 hours. In the same study encouraging results were obtained by designing chitosan and HNT/insulin hybrids for possible patches with transdermal insulin delivery. In this case more relevant results were obtained (approx. 20 hours) (Massaro et al. 2018b).

Laccase, glucose oxidase, and lipase (Tully et al. 2015) were successfully immobilised on HNT by speculating the positively charges on alumina sites, from clay structure. Adsorbed proteins thermal stability was improved by HNT. This stabilisation was reflected also in the optimal temperature for enzyme activity (Tully et al. 2016).

Other polymer modified HNT example was investigated by grafting polyethyleneimine on the inorganic structure (Long et al. 2017). These polymer-HNT structures were able to bind green fluorescence protein. These complexes showed higher transfection efficiency towards both 293T and HeLa cells (compared with neat PEI complexes) (Massaro et al. 2015). This result highlights the promising potential for gene therapy. Promising directions were register also in immunoglobulin delivery (Hartwig et al. 2015).

Cardanol and cardol next to other valuable compounds from cashew nut, were investigated on their biological properties (antimicrobial, anti-inflammatory, antioxidant, and antitumoral). Recent advances highlighted the ability of supramolecular structures based on HNT modified with triazolium salts for loading cardanol (Massaro et al. 2015).

HNT can be loaded with amoxicillin, Brilliant Green, chlorhexidine, doxycycline, gentamicin sulphate, iodine, and potassium clavulanate (Massaro et al. 2015). In all cases a gradual release of the bioactive compound was attained. When HNT were used in combination with electrospun poly-e-caprolactone the biological effect can be improved. By this approach surgical suture or dressing could be further exploited (Patel et al. 2016) (Table 8).

Table 8. The drugs encapsulated in HNT or HNT nanocomposites for controlled release

Role of drug	Drugs	HNT-nanocomposite	References
Restorative dentistry	Chlorhexidine (CHX)	HNT-CHX	Feitosa et al. 2019
Orthodontic bonding	Triclosan (TCN)*	HNT-TCN	Degrazia et al. 2018
Cytotoxicity evaluation	Poly(allylamine) hydrochloride (PAH) poly (diallyldimethhylammonium chloride) (PDADMAC) poly(ethyleneimine) (PEI)	HNT-PAH HNT-PDADMAC HHNT-PEI	Tarasova et al. 2019
Gene delivery	Green fluorescence protein labelled pDNA	PEI-g-HNT	Long et al. 2017
Drug delivery	Sodium salicylate	Sodium salicylate HNT-PVA	Bediako et al. 2018
Cancer therapy	Doxorubicin	HNT-doxorubicin-soybean phospholipid	Li et al. 2018
Cancer therapy	Doxorubicin	HNT-g-chitosan oligosaccharide	Yang et al. 2016
Cancer therapy	Doxorubicin	HNT-poly(ethylene glycol)-acid folic	Wu et al. 2018
Cancer therapy	Curcumin	HNT-g-chitosan	Liu et al. 2016

Role of drug	Drugs	HNT-nanocomposite	References
Cancer therapy	Doxorubicin	HNT-disulphide-β-cyclodextrin-adamantan-poly(ethylene glycol)-acid folic	Hu et al. 2017
Control and sustained release of drug	Norfloxacin	HNT-poly(methacrylic acid or acrylamide-co-ethylene glycol dimethacrylate)	Fizir et al. 2017
Cancer therapy	Doxorubicin	Aur-HNT-Doxorubicin-Bovine serum albumin-acid folic	Zhang et al. 2019
Sustained release of drug	Clotrimazole	HNT-SH, heptakis-6-(tert-butyldimethylsilyl) 2-allyloxy-β-cyclodextrin, cysteamine hydrochloride	Massaro et al. 2018a
Immobilization of enzyme	Lacasse, glucose oxidase, lipase, pepsin	HNT-proteins	Tully et al. 2016
Controlled release of insulin	Insulin	HNT-insulin	Massaro et al. 2018b
Gene delivery	DNA	HNT-g-PEI	Long et al. 2017
Release of Antibacterial Agents	Amoxicillin, chlorhexidine, Brilliant Green, Iodine, Doxycycline	HNT-PCL	Patel et al. 2016
Antigen delivery		Recombinant LipL 32-HNT	Hartwig et al. 2015
Sustained release of drug	Nifedipine (6 wt.%)	(HNT-Nifedipine wt 50%)+MCC wt 45%	Yendluri et al. 2017
Controlled drug release	Silibinin, Curcumin	HNT-Sil-Cur	Massaro et al. 2016a
Release of therapeutic restorative agents	Triclosan (TCN)	HNT(8%w/w)-Triclosan	Cunha et al. 2018
Cancer therapy		HNT-DAS-FA/FITC complex	Grimes et al. 2018
Controlled drug release	Curcumin	HNT-Cur (2.9 wt.% curcumin loading)	Massaro et al. 2016b
Cancer therapy	Binase	HNT-Binase complex (7 wt. % binase loading)	Khodzhaeva et al. 2017
Controlled drug release	BG	HNT-BG-dextrin (as an enzyme-activated tube-end stopper)	Dzamukova et al. 2015.

HNTs in Tissue Engineering

HNT applications in tissue engineering can be noticed especially on scaffolds experimental design (Ji et al. 2017). HNT benefits in gelatine scaffolds applications for example can be depicted on several planes: pores morphology and structure, density and porosity, improvement of thermal stability, stress–strain behaviour (on compression test), ibuprofen slow release. Some small decreases in comparison with the control samples, were observed in cell proliferation (MG63 cells) (Ji et al. 2017). Similar results were obtained also in the case of chitosan-agarose-gelatine hydrogels (Naumenko et al. 2016). In this case next to mechanical properties, biodegradability and biocompatibility was improved by the presence of HNT (*in vitro* and *in vivo* tests) (Naumenko et al. 2016).

Chitosan-alginate with HNT scaffolds were found also very promising candidates for tissue engineering (Afshar and Ghaee 2016). The biocomposite scaffolds were obtained by freeze drying and modified by an amination. HNT presence slightly decreased the water uptake and porosity. The amination process brigs several benefits on cell adherence and proliferation (Afshar et al. 2016). Some synthetic data are shown in table 9.

Table 9. HNT biocomposites for tissue engineering –main obtaining routs

Nanocomposites	Drugs	HNT Loading (%)	Mixing method	Release pattern	References
HNT/Gelatine	Ibuprofen	10-50	Solution casting	Over 100 h	Ji et al. 2017
HNT/chitosan-agarose-gelatine		3, 6	Mixing and freeze-drying		Naumenko et al. 2016
HNT/alkaline-phosphatase		50, 66, 75, 80	Solution mixing		Pietraszek et al. 2019
HNT/chitosan-alginate		8	Solution mixing and freeze-drying		Afshar and Ghaee 2016

Another potential material for scaffold applications was investigated by immobilising alkaline phosphatase (ALP) into HNT nanotubes (Pietraszek et al. 2019). The loading efficiency of ALP was reported 13.5% and the encapsulation efficiency 27%. The biomineralization was highlighted by the formation of "cauliflower-like" hydroxyapatite (Pietraszek et al. 2019).

HNTs in Wound Healing

HNT presence in different biocomposites for wound healing can be explained like in the case of cancer or general pharmaceutical applications, by the ability of the inorganic particles to load different bioactive molecules, in the lumen region and on the surface.

Bacterial infection can be better threaded by using prolonged release of the antibiotics. Such examples considered ciprofloxacin and polymyxin B sulphate loaded in HNT (Shi et al. 2018). A silane coupling was done on HNT in order to improve the lumen adsorption (using vacuum to load the antibiotics). The loaded HNT tubes were incorporated in gelatine based elastomeric nanocomposites (glycerol elastomer using genipin as crosslinker). Successful results were attained against *S. aureus* and *P. aeruginosa* (Shi et al. 2017).

Polycaprolactone/gelatine nanofibers obtained by using the principles of green chemistry were investigated for potential wound healing (Pavliňáková et al. 2018). Different HNTs with similar aspect ratio, (i.e., potential loading capacity per particle) were used. The best mechanical properties were achieved for the composites with 0.5 wt. % HNT. Test on NIH-3T3 mouse fibroblasts highlighted the biocompatibility of the final materials (Pavliňáková et al. 2018).

HNT/chitosan nanocomposites were successfully used in vivo, in murine rat model enhancing the reepithelization (Sandri et al. 2017) after 7 days of treatment.

Vancomycin in HNT alginate gel materials were proven very effective for prolonged release (Kurczewska et al. 2017). Essential for the final system product was the HNT organomodification with APTES (Kurczewska et al. 2017).

Table 10. HNT biocomposites for wound healing – main obtaining routs

Nanocomposites	Drugs	HNT Loading (wt. %)	Mixing method	Release pattern	References
HNT/Glycerol/Gelatine/Ciprofloxacin (3.7/55.6/37/3.7)	Ciprofloxacin, or polymyxin B sulphate	3.7 (modif. with KH550 silane)	Gelling and Melt blending	10 h	Shi et al. 2018
HNT/PCL/Gelatine (three type of HNT)		0.5, 1, 3, 6, 9	Electrospinning		Pavlinakova et al. 2018
HNT/chitosan		98.63-99.99	Solution mixing		Sandri et al. 2017
HNT-APTS/Alginate/Glycerol	Vancomycin hydrochloride		Solution mixing		Kurczewska et al.2017
HNT/PLA	Polydopamine		Extrusion and coating		Wu et al. 2019

Recent approaches considered the 3D printing technology for scaffolds useful also for wound dressing (Wu et al. 2019). Printed polylactic acid was used as pattern after functionalising with dopamine to attach on HNT. This approach allowed cell orientation as proven by *in vitro* trials with human mesenchymal stem cells. By this material design approach the PLA bioadhesion and cell proliferation were improved (Wu et al. 2019). The reported materials could be exploited also in other directions like cell culture scaffolds and biosensors. Some synthetic data are shown in table 10.

Halloysite Nanotubes in Other Applications as Nanocontainers

Halloysite nanotubes have been used as nanocontainers for loading, storage, and controlled release of many chemicals (catalysts, anticorrosion agents, antioxidants, flame retardants, drugs, enzymes, and DNA). Usually, the introduction of chemical agents into the lumen of HNTs was done under vacuum followed by samples washing to remove any unloaded agent (Figure 7).

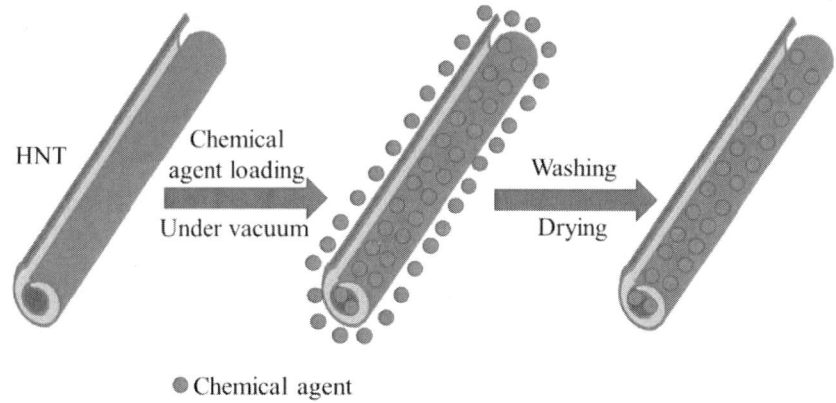

Figure 7. Schematic illustration of chemical agent loading into HNTs lumen.

HNTs in Antifouling

For efficient protection of the surfaces immersed in a marine environment (ships and submarines) against undesirable marine microorganisms proliferation (biofouling) the paint coatings are preferred. The current trend is to develop "green" antifouling agents free of tributyltin (prohibited in 2008 for its toxicity and persistence in the environment) and hazardous substances. Fu et al. 2017 prepared an antifouling thermoplastic composite by mixing the ethylene-vinyl acetate copolymer (EVA) with trichlorophenyl maleimide (TCPM), an antifouling agent with no effect on the environment. TCPM was loaded inside of halloysite clay nanotubes. The EVA-TCPM-halloysite composite allowed the release of the bacterial inhibitor over a period of 12 months versus 3 months if the inhibitor was incorporated directly into the EVA.

HNTs in Agriculture

Another application of HNTs is in agri-food sector. The low use of herbicides, pesticides and fertilisers with increased efficiency, controlled release and targeted delivery are some of the goals of modern agriculture that can be met by using nanotechnology in agriculture. Tan et al. 2015 used HNTs as carriers for the loading and release of the herbicide amitrole (3-amino-1,2,4-trizole, AMT). To increase the loading of AMT, HNTs were modified by first intercalation of dimethyl sulfoxide (DMSO) into the interlayer space of HNTs, followed by mantaining in methanol (MeOH), under stirring for 7 days. The methoxy-modified HNTs exhibited a slow release of AMT because of the significant loading of AMT into the lumen of Halloysite (30.5%). Zeng et al. 2019 have studied the Halloysite nanotubes as nanocarriers for plant herbicide (eupatorium adenophora spreng, AIEAS) and obtaining of biodegradable polymers composite film (poly(vinyl alcohol)/starch composites, with the weight ratio of 80/20) with controlled release of herbicide, used as biodegradable agricultural mulch film. AIEAS loaded in the amount of 7.26 wt. % into the lumen of HNTs was much slower released from PVA/ST film through soil layer than did free AIEAS. Zhong et al. 2017 have studied the loading of herbicide atrazine (AT) into the lumen of HNTs and further dispersion in PVA/ST for obtaining the biodegradable film with controlled release of herbicide. AT loaded into nanotubes in the amount of 9 wt. % displayed much slower release from PVA/ST film in water than free AT (61% after 96 h, compared with 97% in PVA/ST film with free AT).

HNT in Food Packaging

The halloysite nanotubes characteristics such as low cost, high availability, "green" nature and nontoxicity have determined researchers to study HNTs also for possible uses in active food packaging. There is an important interest in the development of natural systems with antibacterial and antiseptic properties and with reduced impact on food. Spepi et al. 2016 have succeeded in loading of 10.5% (w/w) salicylic acid into the HNTs lumen, using HNTs pretreated with H_2SO_4 to enlarge the diameter of the HNT lumen and thus increases the tube loading capacity.

HNTs in Anticorrosion Coating

In last years, researchers have continued to be concerned about the development of self-healing coatings based on inhibitory containers to protect copper. Corrosion inhibitors, encapsulated in microbeads, nanotubes or LDH, could be activated to be released by redox reaction, pH change, ultraviolet radiation, and ionic strength. HNTs due to their nanotubular structure, with lumen of 15 nm and large pore volume, can be used as the storage of anticorrosive agents. Dong et al. 2018 used HNTs as inhibitor nanocontainers and two pH-responsive corrosion inhibitors (L-Valine, L-Val and 2-mercaptobenzothiazole, MBT) were loaded into the lumen of HNTs. The pH-responsive coating composed by pure epoxy, epoxy curing agent, and the composites of L-Val-loaded HNTs and MBT-loaded HNTs was sprayed onto copper substrate. The results proved the higher anticorrosive efficiency of L-Val-loaded HNTs than MBT-loaded HNTs, due to the rapid release performance of L-Val and releases 98% of total content within 300 min.

Halloysite Nanotubes as Flame Retardant

In recent years, one has observed a great interest in polymeric materials with special properties, resistant to both considerably lowered and increased temperatures, flame-retardant, with appropriate mechanical strength. The problem of polymer flammability reduction or making polymers non-flammable acquires an overwhelming importance due to serious health and life hazards as well as environmental pollution caused by the emission of large amounts of smoke and thermal decomposition and combustion products. Huge amounts of heat emitted during fires frequently cause big material losses.

The enhancement of flame retardant properties of nanoparticle reinforced polymer composites have attracted considerable efforts. A way of improving them requires the use of fillers with appropriate properties. The most common flame retardant fillers at nanoscale include inorganic clay minerals (especially magnesium-, silicon-, or aluminium-based). The aluminosilicate clay halloysite nanotubes is an increasingly utilised choice. HNTs generate barriers against heat and mass transport when combustion of the composite material occurs, acting as an insulating layer on the polymer surface (Goda et al. 2018), and also can capture the polymer decomposition products in the lumen.

HNTs can be employed as a flame retardant in PA6, PA11, PA1010, PLA, a high efficiency being obtained with non-toxic and good thermal stability organophosphorus compounds (such as phosphonate and phosphate ester). HNT nanotubes due to their lumen-like structure can be modified with flame retardant by grafting them on interior or exterior surface or by partial encapsulation of flame retardant inside nanotubes. The improved fire retardancy was explained by the slow release of flame retardant and subsequent flame inhibition, together with secondary barrier effect. Smith et al. (2018) succeeded in reducing the flammability of polyurethane foam by using a multilayer nanocoating containing branched polyethylenimine (BPEI) and poly(acrylic acid) (PAA)-stabilised halloysite nanotubes. Only five bilayers of BPEI-HNT/PAA-HNT (≈633 nm thick), deposited from aqueous suspensions, using layer-by-layer assembly, was necessary for self-extinguishing the foam. Loading of bisphenol-A bis(diphenyl phosphate) (BDP) inside the HNT lumen was demonstrated to be an effective method to improve the flame retardancy of polyamide 6 with maintaining mechanical properties. During burning process, the BDP inside the nanotube possessed a prolonged release for flame retardation at the later stage of combustion. The PA6 nanocomposites exhibited the V-0 rating in UL-94 test (Boonkongkaew and Sirisinha 2018). Hao et al. (2014) sudied the flammability properties of PA11. The results showed that PA11, with 25% Clariant's Exolit OP 1312 (Germany), an intumescent flame retardant additive based on metal phosphinate and 2.5% HNT presented the best balance between mechanical and flammability properties (elongation at break of 10.22%, ultimate tensile strength of 30.36 MPa, Young modulus of 1739.86 MPa, Izod impact strength of 5.50 kJ/m^2 and V-0 rating in UL-94 test).

Li et al. (2017) proved the efficiency of DOPO in improving PLA fire retardancy with maintaining thermal stability. HNT nanotubes were sequentially grafted with maleic anhydride (MAH) and DOPO and the nanohybrids resulted were melt blended with PLA. The best results were obtained for PLA composite with 5 % of nanohybrid (LOI of 38.0%, UL-94 vertical burning rating V-0, compared to neat PLA, LOI =24.7% and UL-94 V-2). PHRR was reduced by 20.2% and time to ignition (TTI) was delayed by 10 s. Some synthetic data are shown in table 11.

Halloysite Nanotubes in Environment Protection

Issues related to water pollution and lack of water resources are of major and topical interest, especially in developing countries. The main water pollutants are heavy metal ions, dyes and pesticides. In recent years, HNTs have found applications in the treatment of wastewater including drinking, industrial and domestic wastewater. CNTs are an inexpensive solution for removal of different kinds of heavy metal ions from wastewater, mainly through physical and/or chemisorption mechanisms.

Table 11. Flame retardant nanocomposites

Polymer	HNTs (wt. %)	Coagent (wt. %) [loading efficiency]	Microstructure	% Relative difference in TSR	% Relative difference in peak HRR	Char quality	Reference
PU Foam	5 10	[3 Bilayer HNT (26)] [5 Bilayer HNT (34)]	Good dispersion	48(-) 60(-)	61(-) 62(-)	Porous network remains as hollow shells	Smith et al. 2018
PA6	5, 10	BDP (2 or 4) [16]	Good dispersion	TTI 9(-)	39(-)		Boonkongkaew and Sirisinha 2018
PA11	2.5, 5, 10	Metal phosphinate (25)			27(-)		Hao et al. 2014
PLA	3, 5, 7	DOPO		TTI 17(+)	20(-)		Li et al. 2017

A special attention has been paid to the detection and monitoring of nitrites in drinking water and in environmental samples because of their potential toxicity. Among different analytical techniques, non-enzymatic electrochemical sensors have been successfully applied for the detection of nitrites. Ghanei-Motlagh and Taher (2018) reported an efficient sensor for electrochemical detection of nitrite based on silver/halloysite nanotube/molybdenum disulfide nanocomposite. First, silver nanorods were obtained inside the lumen of the HNTs by chemical method and then the MoS_2 layers have been deposited on the Ag/HNT nanocomposite by hydrothermal method. A carbon paste electrodes (CPEs) was modified with the prepared composite material (Ag/HNT/MoS_2) and was successfully applied as a nitrite electrochemical sensor for environmental applications. The designed sensor was applied for nitrite oxidation.

A major problem faced by the world, due to the increasing industrial development, is the wastewater purification. One of the solutions used in wastewater purification is membrane separation. The main problem with membranes made of polymers (Poly (vinyl chloride) polymer (PVC), polyvinylidene fluoride (PVDF), polyethersulfone (PES) and polysulfone (PS)) is the adsorption of microorganisms or whey on the membrane surface or in the pores and decreasing the water flux. Therefore, many researchers studied the modification of such membranes with hydrophilic additives, by chemical grafting, and surface coating. Mishra and Mukhopadhyay 2018 have found that the antifouling performance of PVC membrane can be enhanced through the uniform incorporation of HNTs. With the new PVC/HNTs membrane the water flux increased almost twice.

The presence of dyes in wastewater causes serious environmental problems, which is why finding efficient removal methods has permanently concerned the researchers. Halloysite nanotubes are negatively charged on the outer surface (negative Si-O-Si) and positive on the surface inside the lumen (positive Al-OH), which is why they can adsorb both anionic and cationic dyes. Desai et al. (2017) used HNTs to remove cationic dyes (like Auramine Y and Auramine O) from aqueous solution. The results have proven the effectiveness of HNTs as an adsorbent for the removal of

Auramine dye. The maximum adsorption capacity was about 64 mg/g, variable with temperature.

The presence of antibiotic residues in the aquatic environment presents a high risk to human health being a class of strong organic pollutants. Many treatment techniques of antibiotics residues have been developed and tested. The most efficiency and easy to be used method is adsorption, although ordinary adsorbents do not have the ability to recognize and to selectively adsorb target species. For this reason, the molecular printing technique (MIT), the surface molecular imprinting technique (SMIT) and hollow molecular imprinting technique were developed. HNTs because of large interior space and excellent and tunable surface properties are widely used as a sacrificial template in all these techniques. Xie et al. (2016) prepared an advanced selective nanoadsorbent, hollow molecularly imprinted nanorods (HMINs), to remove chloramphenicol (CAP). They modified HNTs with 3-(Trimethoxysilyl) propyl methacrylate and then by precipitation polymerization on surface of modified HNTs with CAP (as template molecule) the surface molecular imprinting nanorods (SMINs) were obtained. After the CAP template and HNTs were removed by soxhlet extraction and HF etching, respectively, HMINs were obtained. HMINs presented high specific recognition to CAP, very good thermal stability and regeneration properties, being suitable for using in wastewater treatments.

A major concern, especially in urban areas, was to find an advantageous solution for controlling the quality and quantity of stormwater contaminated with heavy metals from automobiles fluid leaks, tires, paints and atmospheric deposition. Biofiltration systems consist of a layered soil media with different range of particle sizes and are used to remove heavy metal ions. Hermawan et al. (2018) have studied the efficiency of HNTs in removing heavy metal ions Fe(III), Mn(II), Cu(II), Zn(II), Ni(II), and Pb(II), in comparison with fly ash and zeolite. In this regard, three soil columns, of three layers from bottom to top including drainage, transition and filter media were developed. The results proved a higher infiltration rate and a higher heavy metal ions removal capacity for HNTs than for fly ash and zeolite. Filters with 2% HNTs were found to be the best compositions.

HNTs due to their large surface area, biocompatibility, and chemical and mechanical stability are an ideal carrier for immobilisation of enzyme. Fan et al. (2018) proposed two strategies for entrapping and embedding chloroperoxidase (CPO) on both inner/outer wall of HNTs by pH modulated electrostatic adsorption or hydrogen bonding interaction (physical adsorption and chitosan modification). The results have revealed significant improvement of thermal stability and tolerance to organic solvents of immobilised CPO (I-CPO) compared to the free enzyme. Furthermore, I-CPO has proven effective in the degradation of pesticide, isoproturon, indicating the potential application of I-CPO in treatment of wastewater.

Halloysite Nanotubes in Catalysis

Catalysis is the process of increasing the rate of a chemical reaction by adding a substance which is not consumed in the catalysed reaction and which can continue to act several times. Catalysts have a significant importance in various processes such as industrial wastes, environmental remediation, developing renewable source of energy, etc.

Nanomaterials can be very good catalysts because of their high surface area, shape selectivity, microporous structure (Sharma et al. 2015). The microporous structure of HNT can make it to play as a support for various metal-based catalysts.

Also, polymer –supported catalysts are a very important subject. An useful modality to classify this category of catalysts is based on the type of catalysed reactions.

C-C Coupling Reactions

A coupling reaction means the term for the reactions where two fragments are joined together with the aid of a metal catalyst.

First, the HNTs modified with (3-chloropropyl)trimethoxy silane via ultrasonic irradiation was obtained. Then, a unique hybrid system was synthesised based on covalent conjugation of amine-functionalised starch with Cl-HNT, followed by coordination of Pd acetate (Sadjadi et al. 2018a)

(Figure 8). The catalytic activity was evaluated in the copper- and ligand-free Sonogashira coupling reactions. The recyclability was evaluated for ten reaction runs.

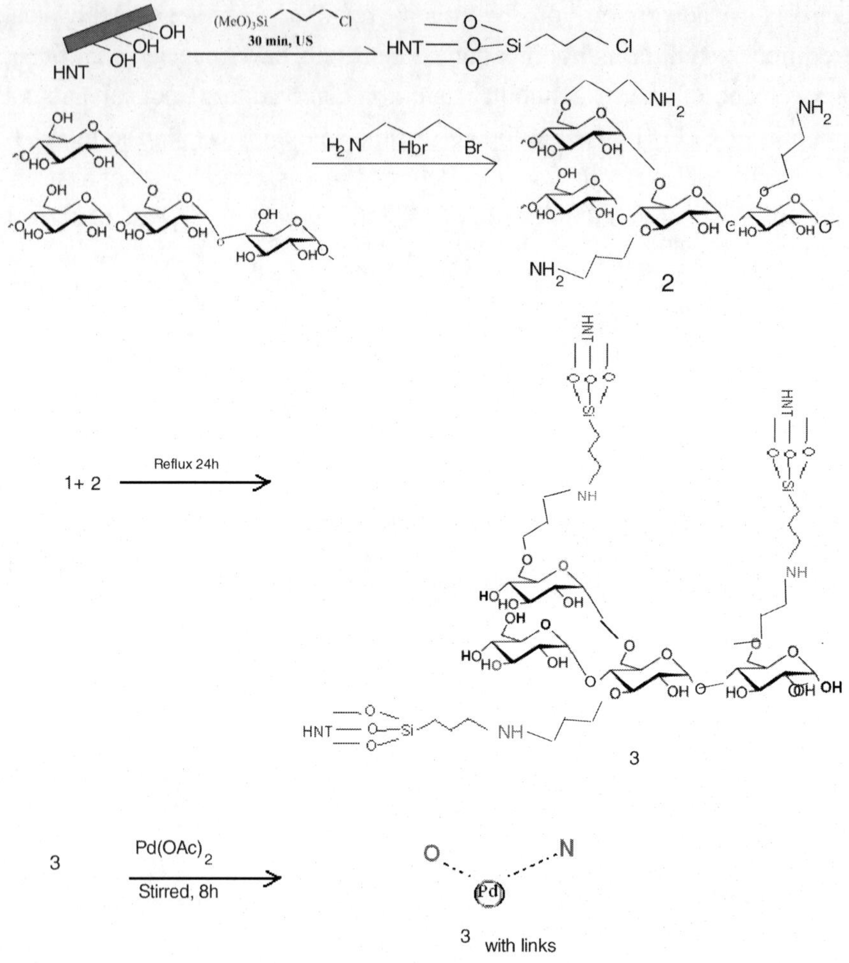

Figure 8. Scheme of reactions (adapted from Sadjadi et al. 2018a).

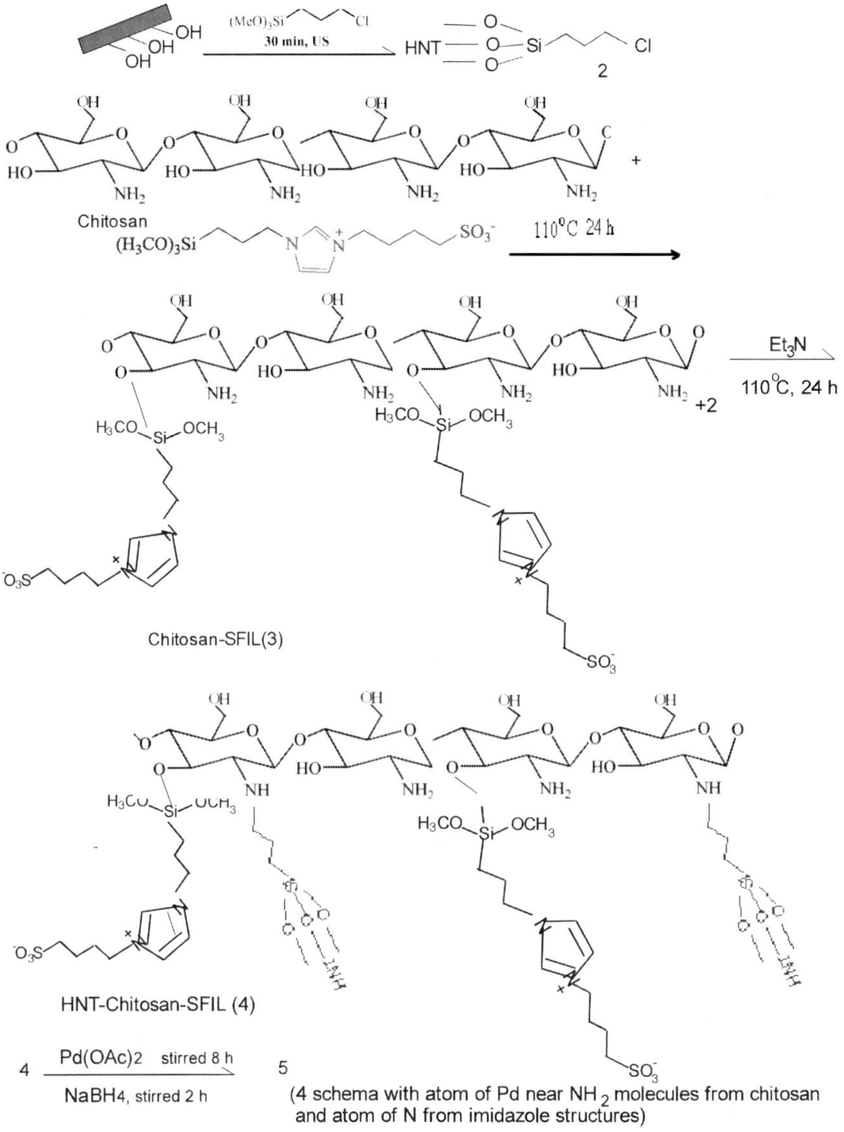

Figure 9. Scheme of reactions (adapted from Sadjadi et al. 2018b).

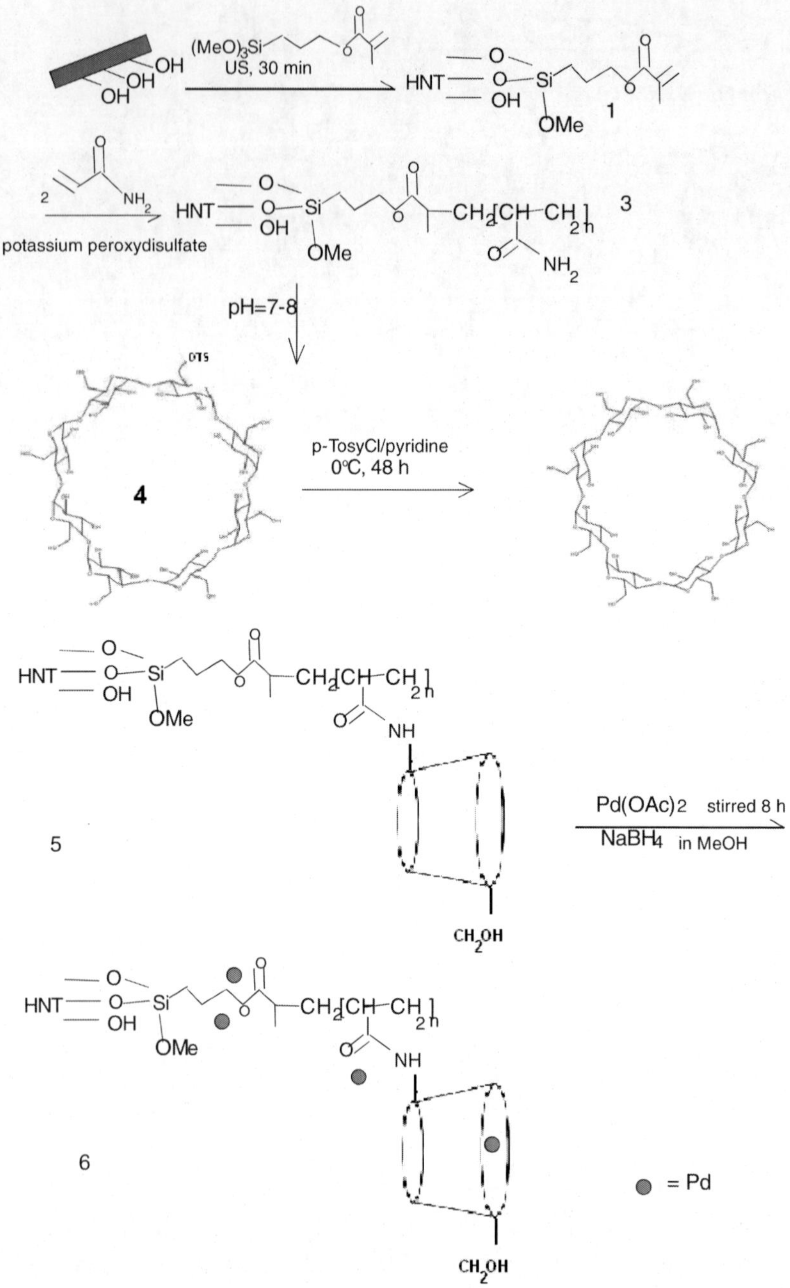

Figure 10. Scheme of reactions (adapted from Sadjadi and Atai 2018).

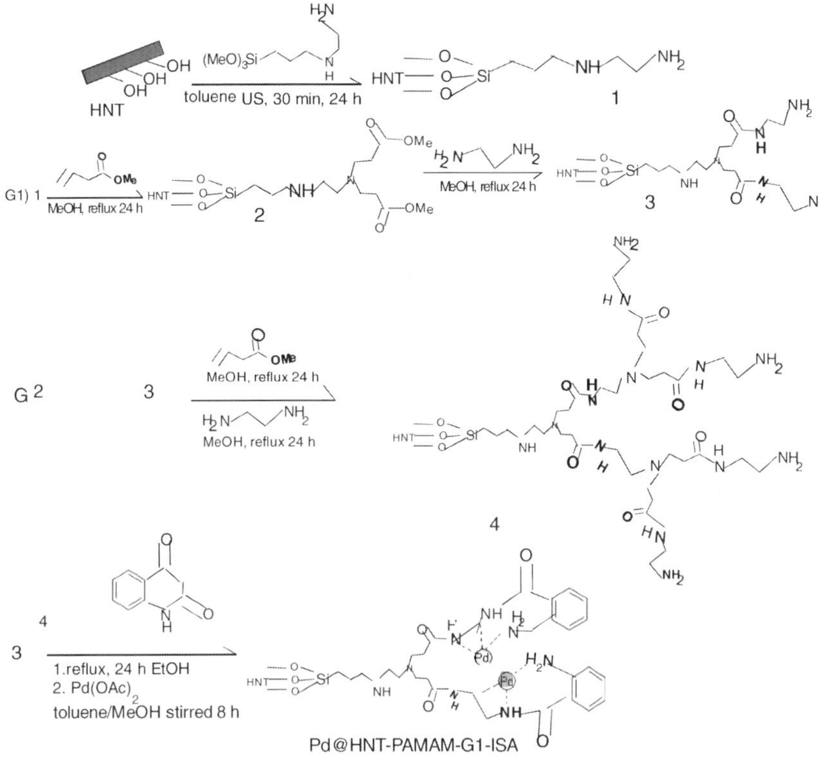

Figure 11. Scheme of reactions (adapted from Bahri-Laleh et al. 2018a).

In addition, the Pd-based halloysite –chitosan-sulfur-functionalised ionic-liquids (SFIL) was prepared via covalent bonding of Cl-functionalised HNT and chitosan-SFIL. The obtained hybrid catalyst was used as a heterogeneous catalyst for catalysing Sonogashira reaction under mild reaction condition. The HNTs modified with (3-chloropropyl)trimethoxy silane via ultrasonic irradiation was obtained. Then a mixture of chitosan and SFIL was prepared and was added to the HNT-Cl. Treatment of HNT-chitosan-SFIL dispersed in toluene, with Pd(OAc)$_2$ followed by reduction with NaBH$_4$ in methanol gave the Pd-based catalyst (Sadjadi et al. 2018b) (Figure 9). The catalytic activity was 96%. The recyclability of the catalyst was examined for seven reaction runs and the loss of catalytic activity after recycling was observed.

HNT as catalyst support was studied by Sadjadi and Atai (2018). A ternary hybrid system was synthesised through growing polyacrylamide (PAA) on the surface of functionalised HNT, followed by introduction of β-cyclodextrin (β-CD). The system was used for immobilisation of Pd nanoparticles and obtaining of a heterogeneous catalyst.

First, HNT was functionalised with 3-(trimethoxysilyl)propyl methacrylate via ultrasonic irradiation. The surface of functionalised HNT was decorated with PAA (using potassium peroxydisulfate as polymerisation initiator). In the same time β-CD was tosylated with p-toluensulfonyl chloride. Then, HNT-PAA-CD was synthesized by reaction of CD-Tosylated, triethylamine and HNT-PAA-CD. To incorporate Pd nanoparticles on HNT-PAA-CD, a palladium acetate was added to HNT-PAA-Cd, and $NaBH_4$ was introduced to allow the reduction of Pd(II) to Pd(0) (Figure 10).

The utility of the catalyst for promoting ligand and copper-free Sonogashira and Heck coupling reaction in aqueous media was evaluated. The recyclability was evaluated for seven reaction runs.

Bahri-Laleh et al. (2018a), reported the developing of an efficient catalyst for copper and ligand free C-C coupling reactions, based on dendrimer decorated HNT preparing.

The starting material was the functionalised HNT with N-[3-(trimethoxysilyl)propyl] ethylenediamine with sonication. Then, the growth of dendrimer on HNT-N was carried out with a two-step procedure, first a Michael addition between methyl acrylate and amino groups of HNT-N, secondly amidation of the terminal ester groups of HNT-N achieved by reaction with ethylenediamine. The obtained hyperbranched polyamidoamine-halloysite (PAMAM)-HNT was subsequently mixed with isatoic anhydride and refluxed 18 h for obtaining HNT-PAMAM-ISA. To incorporate Pd nanoparticles on HNT-PAMAM-ISA, a palladium acetate was added to HNT-PAMAM-ISA, and $NaBH_4$ was introduced to allow the reduction of Pd(II) to Pd(0) (Figure 11). The utility of the catalyst for promoting ligand and copper-free Sonogashira coupling reaction in aqueous media was evaluated. The recyclability was evaluated for ten consecutive reaction runs.

Photocatalytic Reactions

Photocatalysis encompasses a class of reactions which use a substance named catalyst activated by light. These reactions include the decomposition of organic compounds into water and carbon dioxide, leading to exciting properties of surfaces covered with a photocatalyst.

A very interesting and comprehensive review about HNT-based nanocomposites and heterogeneous catalysis was reported by Papoulis (2019). The photocatalytic activities of HNT-based nanocomposites in combination with various photocatalysts that have been synthesised and tested in decomposing air and organic pollutants [such as, antibiotics (Li et al. 2015, Li et al. 2016, Wu et al. 2018), aniline and its derivatives (Szczepancik et al. 2017), azo dyes (Fatimah et al. 2018, Zheng et al. 2016), other organic pollutants (Wang et al. 2018)] was reviewed. To avoid unnecessary redundancy with the above mentioned review, the selected papers have not been presented in detail.

Miscellaneous Examples

Bahri-Laleh et al. (2018b), reported a heterogeneous catalyst based on imine-functionalisation of HNTs. The starting material was the amine-functionalised HNTs. The HNTs modified with N-[3(trimethoxysilyl) propyl] ethylenediamine were obtained, followed by condensation with salicylaldehyde. The resulting functionalised HNTs were applied for the immobilisation of CuI (Figure 11). The latter was used for promoting click reactions of terminal alkynes, sodium azide and α-haloketones in aqueous media and under mild reaction conditions to obtain 1,2,3-triazides in short reaction times. The catalytic activity was 98%. The recyclability of the catalyst was examined for six reaction runs.

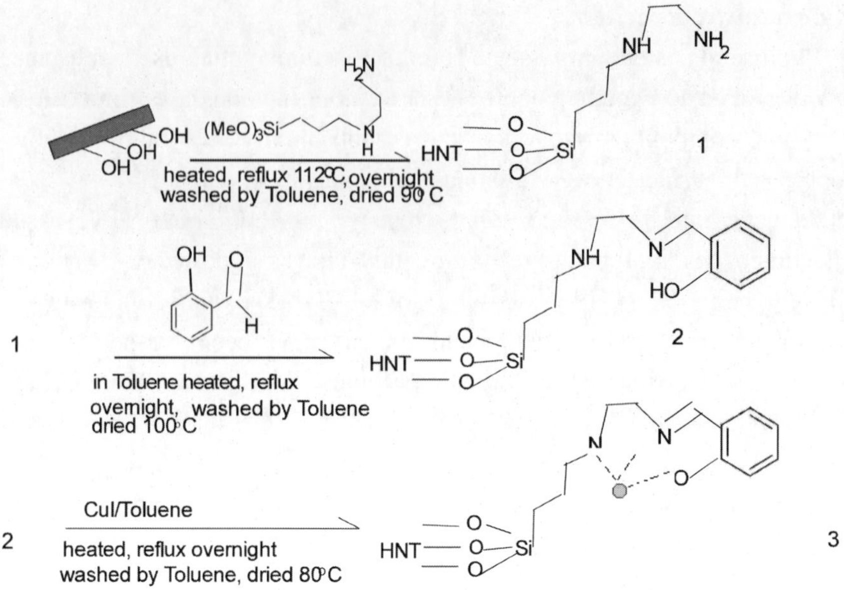

Figure 12. Scheme of reactions (adapted from Bahri-Laleh et al. 2018b).

Maleki et al. (2017) reported the design, preparation and characterisation of a novel superparamagnetic heterogeneous nanocatalyst Fe_3O_4-HNT-SO_3H. First, preparation of Fe_3O_4-HNT was achieved by adding HNT to a solution of $FeCl_3.6H_2O$ and $FeCl_2.4H_2O$ (in a ratio 73:27) under N_2 and 60°C. To prepare iron oxide, a 8 mol/l solution of (NH_3, H_2O) was added to the above mixture, with controlling the pH in the range of 9-10.

Secondly, Fe_3O_4-HNT was dispersed in CH_2Cl_2 by ultrasonication, afterward a solution of chlorosulfonic acid in CH2Cl2 was added with stirring at 0°C. Finally, the catalyst was separated from the reaction mixture by a magnetic bar. The catalyst was used in synthesising of dihydropyrimidinones a-n. The recyclability was evaluated for ten consecutive reaction runs.

The synthesis by chemical reduction of a nanocatalyst based on HNTs and Au nanoparticles (HNT@Au) was reported by Massaro et al. (2019). First, a thiol-functionalized HNT was prepared, by adding 3-mercaptopropyltrimethoxysilane to a weighed amount of HNT under ultrasonication.

Secondly, thiol-functionalized HNT, $HAuCl_4.3H_2O$ and water were stirred for 18 h at room temperature, filtered and washed. The obtained material was re-suspended in MeOH, and was added $NaBH_4$ as a reducing agent under stirring. The catalytic performance of the HNT@Au hybrid nanomaterial was evaluated by the reduction of 4-nitrophenol to 4-aminophenol. The recyclability was evaluated for three consecutive reaction runs. Tsoufis et al. (2017), reported the in situ synthesis of relatively small magnetite nanoparticles dispersed along the surface of HNT by a modified wet-impregnation method. The methodology includes the immobilisation of the iron cations at the surface of HNT, followed by interaction with acetic acid vapours and calcinations. HNT was dispersed in CH_3OH containing $Fe(NO_3)_3.9H_2O$ to yield samples containing 17% wt iron, by stirring 18 hours. The powder obtained after removal of the solvent was exposed to vapours of acetic acid at 80°C, 1 h and dried. The nanocatalyst was obtained by calcinations for 1 h at 200°C under argon flow. The catalytic performance was evaluated by decomposition of a pentachlorophenol solution in CH_3OH, in presence of $NaIO_4$ oxidant. The recyclability was evaluated for three consecutive reaction runs.

A simple one-pot synthetic method used to immobilise Rh, Pd, or Pt nanoparticles onto the surface of the HNT was reported by Jaine and Mucalo 2019.

HNT was suspended in solutions of either $RhCl_3$, Na_2PdCl_4, or H_2PtCl_6. Concentrations of the metal salts were set such that in 100 ml of solution there was 30 mg of each metal, so that the final catalysts would have a loading of 1.0 wt. %. Reduction of the metal salts to their finely metallic form was brought about by addition of 0.05 mol L^{-1} $NaBH_4$ solution. After that the suspensions were stirred for 30 min and a 0.1 mol L^{-1} $LaCl_3$ solution was added to destabilise any material remaining in colloidal suspension. The materials were filtered, washed, dried at 100°C, and stored under vacuum.

The catalytic activity of the materials was tested using olefin hydrogenation as a probe reaction. The olefins were cyclohexene, 1,5-cyclooctadiene, and 1-octene.

In the table 12 some synthetic data from the above mentioned papers are presented.

Table 12. Examples of HNT applications in catalysis

Catalysts	Hybrids	Catalyst amount	Hydro-dynamic radius (nm)	Time (min)	Yield (%)	References
Pd@HNT-(bromoethyl amine-functionalised starch)	HNT-(bromoethyl amine-functionalised starch)			75	95	Sadjadi et al. 2018a
Pd@HNT-Chitosan-Sulfur functionalised Ionic Liquids (SFILs)	HNT-Chitosan-SFILs				96	Sadjadi et al. 2018b
Pd@HNT-Polyacrylamide-Cyclodextrin				75	95	Sadjadi and Atai 2018
Pd@HNT-PAMAM-G1-ISA; Pd@HNT-PAMAM-G2-ISA				75 75	80 68	Bahri-Laleh et al. 2018a
CuI@HNTs - 2 N - Sal		0.48 mol%		12	98	Bahri-Laleh et al. 2018b
Fe_3O_4-HNT-SO_3H		0.05 g 0.06 g			96 97	Maleki et al. 2018
HNT@Au			395			Massaro et al. 2019

CONCLUSION

In this chapter we provided an overview of the researches foccusing on the last 4 years on the applications of HNT, in different fields and topics like biomedicine and pharmacotherapy, food packaging, agriculture, anticorrosion coating, water treatment, catalysis, antifouling, or flame retardant. A special attention was given to HNT as reinforcement for polymer composites (based on different polymer matrix like PA6, PA11, PMMA, PLA, PS, PP, PVA, EVA, PHB, PEG, thermoplastic starch, or CMC).

Due to its special properties, such as high mechanical strength, chemical resistance, nanotubular structure, with lumen of 15 nm, high aspect ratio, low cost, high availability, "green" nature and non toxicity, HNT has found applications in different fields.

The incorporation of HNT in thermoplastic polymers has allowed the obtaining of nanocomposites with improved mechanical, thermal and barrier properties against gases (e.g., O_2 and CO_2) and water vapours. For compatibilisation of hydrophobic polymers with hydrophilic HNT, grafted polymers and coupling agents (organosilanes) were used. Thus, the adhesion to the polymer-HNT interface became stronger and, consequently, the dispersion of HNT, thermal stability and mechanical properties increased.

The trend of recent years is to replace the synthetic plastics, where is possible, with natural polymer materials due to their low cost, biocompatibility, and biodegradability. However, they are inflexible, hydrophilic and with poor mechanical properties. The mechanical, thermal and barrier properties were improved by incorporation of inorganic nanoparticles such as HNTs.

Numerous studies have shown that HNTs have the capacity of loading multiple active agents simultaneously, are cytocompatible and biocompatible. Due to these properties HNTs can be used as drug delivery systems for active substance delivery to tissues and organs in the body, that are infected or diseased. Therefore, HNTs found applications in cancer therapy, in pharmaceutical, in tissue engineering and in wound healing.

By simple modification of both inside and outside surfaces of HNT interesting properties requested by different applications were obtained. HNTs have been used as nanocontainers for loading, storage, and controlled release of many chemical agents (catalysts, anticorrosion and antifouling agents, antioxidants, flame retardants, drugs, enzymes, and DNA). The chemical agents were introduced into the lumen of HNT under vacuum.

The "green" antifouling agents, free of tributyltin, were prepared by HNTs modification with bacterial inhibitors, with no effect on the environment. The antifouling agents were loaded inside of HNTs and were released controlled for a certain period of time from thermoplastic composites.

For application in agri-food sector, HNTs were used as carriers for the loading and release of herbicides, pesticides and fertilisers and further obtaining of biodegradable polymers composite film.

HNTs were studied for uses in active food packaging. For this purpose, natural substances with bactericidal and antiseptic properties were loaded into the HNTs lumen.

HNTs can be used as inhibitor nanocontainers to protect copper. Corrosion inhibitors, loaded in HNTs, could be activated to be released by redox reaction, pH change, ultraviolet radiation, and ionic strength.

HNTs can be employed as a flame retardant in PA6, PA11, PA1010, PLA and other thermoplastics. HNTs can be modified with flame retardant by grafting them on interior or exterior surface or by partial encapsulation of flame retardant inside nanotubes. The improved fire retardancy was explained by the slow release of flame retardant and subsequent flame inhibition, together with secondary barrier effect.

HNTs have found applications in the treatment of wastewater including drinking, industrial and domestic wastewater. HNTs are an inexpensive solution for removal of different kinds of heavy metal ions, some dyes or some organic pollutants from wastewater, mainly through physical and/or chemisorption mechanisms.

HNTs can be very good catalysts because of their high surface area, shape selectivity, microporous structure. Many researchers studied HNT as catalyst support for immobilisation of Pd nanoparticles and obtaining of a heterogeneous catalyst or as an efficient catalyst for copper and ligand free C-C coupling reactions as well as for obtaining of a novel superparamagnetic heterogeneous nanocatalyst. Also, the photocatalytic activities of HNT-based nanocomposites in combination with various photocatalysts have been studied.

Overall, HNTs are cheap materials with interesting features for many applications. Due to the ease with which they can be modified / functionalised, they will be in the future a good subject of study for obtaining new polymeric nanocomposites with multiple uses.

ACKNOWLEDGMENTS

This work was supported by a grant of the Romanian Ministry of Research and Innovation, CCCDI - UEFISCDI, project number PN-III-P1-1.2-PCCDI-2017-0387/80PCCDI "Emerging technologies for the industrial use of 2 D structures (graphene and non-graphene)" Acronym EMERG2Ind, within PNCDI III.

REFERENCES

Abdallah, R. M. 2016. "Evaluation of polymethyl methacrylate resin mechanical properties with incorporated halloysite nanotubes." *The Journal of Advanced Prosthodontics* 8:167-171.

Abdullah, Z. W. and Dong, Y. 2018. "Preparation and characterisation of poly(vinyl) alcohol (PVA)/starch (ST)/halloysite nanotube (HNT) nanocomposite films as renewable materials." *Journal of Materials Science* 53:3455-3469.

Afshar, H. and A., Ghaee, A. 2016. "Preparation of aminated chitosan/alginate scaffold containing halloysite nanotubes with improved cell attachment." *Carbohydrate Polymers* 151:1120-1131.

Alakrach, A. M., Noriman, N. Z., Dahham, O. S., Hamzah, R., Alsaadi, M. A., Shayfull, Z. and Syed Idrus, S. Z. 2018. „The Effects of Tensile Properties of PLA/HNTs- ZrO$_2$ Bionanocomposites." *IOP Journal of Physics: Conference Series* 1019:012066.

Ali, U., Abd K., Khairil J. and Buang, N. A. 2015. "A Review of the Properties and Applications of Poly (Methyl Methacrylate)." *Polymer Reviews* 55:678-705.

Aloui, H., Khwaldia, K., Hamdi, M., Fortunati, E., Kenny, J. M., Buonocore, G. G. and Lavorgna, M.. 2016. "Synergistic Effect of Halloysite and Cellulose Nanocrystals on the Functional Properties of PVA Based Nanocomposites." *ACS Sustainable Chemistry & Engineering* 4:794-800.

Arat, R. and Uyanik, N. 2017. "Study of the morphological and thermal properties of polystyrene nanocomposites based on modified halloysite nanotubes with styrene-maleic anhydride copolymers." *Materials Today Communications* 13:255-262.

Avella, M., Martuscelli, E. and Raimo, M. 2000. "Review Properties of blends and composites based on poly(3-hydroxy) butyrate (PHB) and poly (3-hydroxybutyrate-hydroxyvalerate) (PHBV) copolymers." *Journal of Materials Science* 35:523-545.

Bahri-Laleh, N., Sadjadi, S., and Poater, A. 2018a. "Pd immobilized on dendrimer decorated halloysite clay: Computational and experimental study on the effect of dendrimer generation, Pd valence and incorporation of terminal functionality on the catalytic activity." *Journal of Colloid and Interface Science* 531:421-432.

Bahri-Laleh, N., Sadjadi, S., Geravi, M. M. and Malmir, M. 2018b. "CuI-functionalized halloysite nanoclay as an efficient heterogeneous catalyst for promoting click reactions: Combination of experimental and computational chemistry." *Applied Organometallic Chemistry* 32:e4283.

Bediako, E. G., Nyankson, E., Dodoo-Arhin, D., Agyei-Tuffour, B., Lukowiec, D., Tomiczek, B., Yaya, A. and Efavi, J. K. 2018. "Modified halloysite nanoclay as a vehicle for sustained drug delivery." *Heliyon* 4: https://doi.org/10.1016/j.heliyon.2018.e00689.

Beltran, F. R., de la Orden, M. U. and Urreaga, J. M. 2018. "Amino-Modified Halloysite Nanotubes to Reduce Polymer Degradation and Improve the Performance of Mechanically Recycled Poly(lactic acid)." *Journal of Polymers and the Environment* 26:4046–4055.

Bettencourt, A. and Almeida, A. J. 2012. "Poly(methyl methacrylate) particulate carriers in drug delivery." *Journal of Microencapsulation Micro and Nano Carriers* 29:353-367.

Bidsorkhi, H. C., Adelnia, H., Pour, R. H. and Soheilmoghaddam M. 2015 "Preparation and characterization of ethylene-vinyl acetate/halloysite nanotube nanocomposites."*Journal of Materials Science* 50:3237-3245.

Boonkongkaew, M. and Sirisinha, K. 2018. "Halloysite nanotubes loaded with liquid organophosphate for enhanced flame retardancy and mechanical properties of polyamide 6." *Journal of Materials Science* 53:10181-10193.

Borggreve, R. J. M. and Gaymans, R. J. 1989. "Impact behaviour of nylon-rubber blends 4. Effect of the coupling agent, maleic anhydride." *Polymer* 30:63-70.

Bucknall, C. B.and Paul, D. R. 2009. "Notched impact behavior of polymer blends: part 1: new model for particle size dependence." *Polymer* 50:5539-5548.

Bugatti, V., Sorrentino, A. and Gorrasi G. 2017. "Encapsulation of Lysozyme into halloysite Nanotubes and dispersion in PLA: Structural and physical properties and controlled release analysis." *European Polymer Journal* 93:495-506.

Buruga, K., Kalathi, J. T., Kim, K.-H., Ok, Y. S. and Boukhvalov, D. 2018. "Polystyrene-halloysite nano tube membranes for water purification" *Journal of Industrial and Engineering Chemistry* 61:169-180.

Cermak, M., Kadlec, P., Sutta, P. and Polansky, R. 2016. "Structural and mechanical behaviour of LLDPE/HNT nanocomposite films." *AIP Conference Proceedings* 1713, 090006-1-090006-5.

Chen, Y., Murphy, A., Scholz, D., Geever, L. M., Lyons, J. G. and Devine D. M. 2018. "Surface-modified halloysite nanotubes reinforced poly (lactic acid) for use in biodegradable coronary stents." *Journal of Applied Polymer Science* 135: https://doi.org/10.1002/app.46521.

Chen, Y., Geever, L. M., Killion, J. A., Lyons, J. G., Higginbotham, C. L. and Devine, D. M. 2017. "Halloysite Nanotube Reinforced Polylactic Acid Composite." *Polymer Composites* 38:2166-2173.

Cheng, Z.-L., Ma, L. and Liu, Z. 2018a "A study on synergistic reinforcing effect of halloysite nanotubes/diatomite mixture-filled polymer (PP and PA6) composites." *Plastics, Rubber and Composites* 47:1-9.

Cheng, Z.-L., Chang, X.-Y., Liu, Z. and Qin, D.-Z. 2018b. "Surface-modified halloysite nanotubes as fillers applied in reinforcing the performance of polytetrafluoroethylene." *Clay Minerals* 53:643-656.

Cheng, Z.-L., Qin, X.-X., Liu, Z. and Qin, D.-Z. 2017a. "Electrospinning preparation and mechanical properties of PVA/HNTs composite nanofibers."*Polymers for Advanced Technologies*, 28:768–774.

Choo, C. K., Kong, X. Y., Goh, T. L., Ngoh, G. C., Horri, B. A. and Salamatinia, B. 2016. „Chitosan/halloysite beads fabricated by ultrasonic-assisted extrusion-dripping and a case study application for copper ion removal." *Carbohydrate Polymers* 138:16-26.

Chow, W. S., Tham, W. L., Poh, B. T. and Mohd Ishak Z. A. 2018. "Mechanical and Thermal Oxidation Behavior of Poly(Lactic Acid)/Halloysite Nanotube Nanocomposites Containing N,N'-Ethylenebis(Stearamide) and SEBS-g-MA." *Journal of Polymers and the Environment* 26:2973-2982.

Cunha, D. A., Rodrigues, N. S., Souza, L. C., Lomonaco, D., Rodrigues, F. P., Degrazia, F. W., Collares, F. M., Sauro, S. and Saboia V. P. A. 2018. "Physicochemical and Microbiological Assessment of an Experimental Composite Doped with Triclosan- Loaded Halloysite Nanotubes." *Materials* 11:1080

Degrazia, F. W., Genari, B., Leitune, V. C. B., Arthur, R. A., Luxan, S. A., Samuel, S. M. W., Collares, F. M. and Sauro, S. 2018. "Polymerisation, antibacterial and bioactivity properties of experimental orthodontic adhesives containing triclosan-loaded halloysite nanotubes." *Journal of Dentistry* 69:77-82.

Desai, S., Pandey, A. and Dahiya M. S. 2017. "Application of Hallosysite Nanotubes in Removal of Auramine Y and Auramine O Dyes." *International Journal of PharmTech Research* 10:62-76.

De Silva, R. T., Soheilmoghaddam, M., Goh, K. L., Wahit, M. U., Abd Hamid, S. B., Chai, S.-P. and Pasbakhsh, P. 2016. "Influence of the Processing Methods on the Properties of Poly(lactic acid)/Halloysite Nanocomposites." *Polymer Composites* 37:861-869.

Dong, C., Zhang, M., Xiang, T., Yang, L., Chan, W. and Li C. 2018. "Novel self-healing anticorrosion coating based on L-valine and MBT-loaded halloysite nanotubes." *Journal of Materials Science* 53:7793-7808.

Dulebova, L., Glogowska, K., Hajek, J. and Fic, J. 2018. "The effect of adding halloysite nanotubes as filler on the mechanical properties of low-density polyethylene." *Materials Science Forum* 919:144-151.

Dzamukova, M. R., Naumenko, E. A., Lvov, Y. M. and Fakhrullin, R. F. 2015. "Enzyme-activated intracellular drug delivery with tubule clay nanoformulation." *Scientific Reports* 5:10560.

Fan, X., Hu, M., Li, S., Zhai, Q., Wang, F. and Jiang, Y. 2018. "Charge controlled immobilization of chloroperoxidase on both inner/outer wall of NHT: Improved stability and catalytic performance in the degradation of pesticide." *Applied Clay Science* 163:92-99.

Fatimah, I. S. and Herianto, R.. 2018. "Physicochemical Characteristics and Photocatalytic Activity of Silver Nanoparticles-decorated on Natural Halloysite (An aluminosilicate clay)." *Oriental Journal of Chemistry* 34:857-862.

Feitosa, S. A., Palasuk, J., Geraldeli, S., Windsor, L. J. and Bottino M. C. 2019. "Physicochemical and biological properties of novel chlorhexidine-loaded nanotube-modified dentin adhesive." *Journal of Biomedical Materials Research B Part B*. 107:868-875.

Fizir, M., Dramou, P., Zhang, K., Sun, C., Pham-Huy, C. and He, H. 2017. "Polymer grafted-magnetic halloysite nanotube for controlled and sustained release of cationic rug." *Journal of Colloid and Interface Science* 505:476-488.

Fu, Y., Gong, C., Wang, W., Zhang, L., Ivanov, E. and Lvov, Y. 2017. "Antifouling Thermoplastic Composites with Maleimide Encapsulated in Clay Nanotubes." *ACS Applied Materials and Interfaces* 9:30083-30091.

Fukushima, K., Wu, M.-H., Bocchini, S., Rasyida, A. and Yang, M.- C. 2012. "PBAT based nanocomposites for medical and industrial applications." *Materials Science and Engineering: C* 32:1331-1351.

Gaaz, T. S. and Hussein, E. K. 2017. "Physical Properties of Halloysite Nanotubes- Polyvinyl Alcohol Nanocomposites Using Malonic Acid Crosslinked." *Jurnal Kejuruteraan* 29:71-77.

Gaaz, T. S., Sulong, A. B., Akhtar, M. N., Kadhum, A. A. K., Mohamad, A. B. and Al-Amiery, A. A. 2015. "Properties and Applications of Polyvinyl Alcohol, Halloysite Nanotubes and Their Nanocomposites." *Molecules* 20:22833-22847.

Gamini, S., Vasu, V. and Bose, S. 2017. "Tube-like natural halloysite/ poly(tetrafluoroethylene) nanocomposites: simultaneous enhancement in thermal and mechanical properties." *Materials Research Express* 4:045301.

Garcia-Garcia, D., Garcia-Sanoguera, D., Fombuena, V., Lopez-Martinez, J. and Balart R. 2018. "Improvement of mechanical and thermal properties of poly(3- hydroxybutyrate) (PHB) blends with surface-modified halloysite nanotubes (HNT)." *Applied Clay Science* 162:487-498.

Ghanei-Motlagh, M. and Ali Taher, M. 2018. "A novel electrochemical sensor based on silver/halloysite nanotube/molybdenum disulfide nanocomposite for efficient nitrite sensing." *Biosensors and Bioelectronics* 109:279-285.

Goda, E. S., Yoon, K. R., El-sayed, S. H. and Hong, S. E. 2018. "Halloysite nanotubes as smart flame retardant and economic reinforcing materials: A review." *Thermochimica Acta* 669:173-184.

Grimes, W. R., Luo, Y., McFarland Jr., Antwine W. and Mils, D. K. 2018. "Bi-Functionalized Clay_Nanotubes for Anti-Cancer Therapy." *Applied Sciences* 8:281.

Guo, J., Qiao, J. and Zhang, X. 2016. "Effect of an alkalized-modified halloysite on PLA crystallization, morphology, mechanical, and thermal properties of PLA/halloysite nanocomposites." *Journal of Applied Polymer Science* 133:6519-6527.

Handge, U. A., Hedicke-Hochstotter, K. and Altstadt, V.. 2010. "Composites of polyamide 6 and silicate nanotubes of the mineral halloysite: Influence of molecular weight on thermal, mechanical and rheological properties." *Polymer* 51:2690-2699.

Hao, A., Wong, I., Wu, H., Lisco, B., Ong, B., Sallean, A., Butler, S., Londa, M. and Koo, J. H. 2015. "Mechanical, thermal, and flame-retardant performance of polyamide 11–halloysite nanotube nanocomposites." *Journal of Materials Science* 50:157-167.

Hartwig, D. D., Bacelo, K. I., Oliveira, T. I., Schuch, R., Seixas, F. K., Collares, T., Rodrigues, O., Hartleben, C. P. and Dellagostin, O. A. 2015. "The use of halloysite clay and carboxyl-functionalised multi-walled carbon nanotubes for recombinant LipL32 antigen delivery enhanced the IgG response." *Memorias Do Instituto Oswaldo Cruz* 110:134-137.

Hermawan, A. A., Chang, J. W., Pasbakhsh, P., Hart, F. and Talei, A. 2018. "Halloysite nanotubes as a fine grained material for heavy metal ions removal in tropical biofiltration systems." *Applied Clay Science* 160:106-115.

Hu, Y., Chen, J., Li, X., Sun, Y., Huang, S., Li, Y., Liu, H., Xu, J. and Zhong, S. 2017. "Multifunctional halloysite nanotubes for targeted delivery and controlled release of doxorubicin *in-vitro* and *in-vivo* studies." *Nanotechnology* 28:375101.

Huang, B., Liu, M. and Zhou, C. 2017. "Chitosan composite hydrogels reinforced with natural clay nanotubes." *Carbohydrate Polymers* 175:689-698.

Jaine, J. E. and Mucalo, M. R. 2019. "Synthesis, characterisation, and catalytic properties of halloysite supported metal nanoparticles." *Materials Research Bulletin* 111:251-258.

Jenifer, A., Rasana, N. and Jayanarayanan, K. 2018. "Synergistic effect of the inclusion of glass fibers and halloysite nanotubes on the static and dynamic mechanical, thermal and flame retardant properties of polypropylene." *Materials Research Express* 5:065308 https://doi.org/10.1088/2053-1591/aac67d.

Ji, L., Qiao, W., Zhang, Y., Wu, H., Miao, S., Cheng, Z., Gong, Q., Liang, J. and Zhu, A. 2017. "A gelatin composite scaffold strengthened by drug-loaded halloysite nanotubes." *Materials Science and Engineering C* 78:362-369.

Kaygusuz, I. and Kaynak C. 2015. "Influences of hallosite nanotubes on crystallization behaviour of polylactide." *Plastics, Rubber and Composites* 44:41-49.

Kennouche, S., Le Moigne, N., Kaci, M., Quantin, J.-C., Caro-Bretelle, A.-S., Delaite, C. and Lopez-Cuesta, J.-M. 2016. "Morphological characterization and thermal properties of compatibilized poly(3-hydroxybutyrate-co- hydroxyvalerate) (PHBV)/poly(butylenes succinate) PBS)/halloysite ternary nanocomposites." *European Polymer Journal* 75:142-162.

Khodzhaeva, V., Makeeva, A., Ulyanova, V., Zelenikhin, P., Evtugyn, V., Hardt, M., Rozhina, E., Lvov, Y., Fakhrullin, R. and Ilinskaya, O. 2017. "Binase Immobilized on Halloysite Nanotubes Exerts Enhanced Cytotoxicity toward Human Colon Adenocarcinoma Cells." *Frontiers in Pharmacology* 8:631.

Kim, Y. H., Kwon, S. H., Choi, H. J., Choi, K., Kao, N., Bhattacharya, S. N. and Gupta R. K. 2016. "Thermal, Mechanical, and Rheological Characterization of Polylactic acid/Halloysite Nanotube Nanocomposites." *Journal of Macromolecular Science, Part B*. 55:680-692.

Krishnaiah, P., Ratnam, C. T. and Manickam, S. 2017. "Development of silane grafted halloysite nanotube reinforced polylactide nanocomposites for the enhancement of mechanical, thermal and dynamic-mechanical properties." *Applied Clay Science* 135:583-595.

Kubade, P. and Tambe, P. 2017. "Influence of surface modification of halloysite nanotubes and its localization in PP phase on mechanical and thermal properties of PP/ABS blends." *Composite Interfaces* 24:469-487.

Kurczewska, J., Pecyna, P., Ratajczak, M., Gajecka, M. and Schroeder, G. 2017. "Halloysite nanotubes as carriers of vancomycin in alginate-based wound dressing." *Saudi Pharmaceutical Journal* 25:911-920.

Li, J., Zhou, M., Ye, Z., Wang, H., Ma, C., Huo, P. and Yan, Y. 2015. "Enhanced photocatalytic activity of g-C_3N_4-ZnO/HNTs composite heterostructure photocatalysts for degradation of tetracycline under visible light irradiation." *RSC Advances* 5:91177-91189.

Li, X., Zhu, W., Yan, X., Lu, X., Yao, C. and Ni, C. 2016. "Hierarchical $La_{0.7}Ce_{0.3}FeO_3$/halloysite nanocomposite for photocatalytic degradation of antibiotics." *Applied Physics A Materials Science & Processing* 122, 723.

Li, X., Yang, Q., Ouyang, J., Yang, H. and Chang, S. 2016. "Chitosan modified halloysite nanotubes as emerging porous microspheres for drug carrier." *Applied Clay Science* 126:306-312.

Li, Z., Exposito, D. F., Gonzalez, A. J. and Wang, D.-Y. 2017. "Natural halloysite nanotube based functionalized nanohybrid assembled via phosphorus-containing slow release method: A highly efficient way to impart flame retardancy to polylactide." *European Polymer Journal* 93:458-470.

Li, X., Tan, D., Xie, L., Sun, H., Sun, S., Zhong, G. and Ren, P. 2018. "Effect of surface property of halloysite on the crystallization behavior of PBAT." *Applied Clay Science* 157:218-226.

Li, K., Zhang, Y., Chen, M., Hu, Y., Jiang, W., Zhou, L., Li, S., Xu, M., Zhao, Q. and Wan R. 2018. "Enhanced antitumor efficacy of doxorubicin-encapsulated halloysite nanotubes." *International Journal of Nanomedicine* 13:19-30.

Liu, M., Chang, Y., Yang, J., You, Y., He, R., Chen, T. and Zhou, C. 2016. "Functionalized halloysite nanotube by chitosan grafting for drug delivery of curcumin to achieve enhanced anticancer efficacy." *Journal of Materials Chemistry B* 4:2253- 2263.

Liu, Z., Li, Y., Ma, L., Qin, D., Cheng, Z. 2017. "A Study on Tribological Properties of Polypropylene Nanocomposites Reinforced with Pretreated HNTs." *China Petroleum Processing and Petrochemical Technology* 19:115-122.Liu, H., Wang, Z.-G., Liu, S.-L., Yao, X., Chen, Y., Shen, S., Wu, Y. and Tian, W. 2019. "Intracellular pathway of halloysite nanotubes: potential application for antitumor drug delivery." *Journal of Materials Science* 54:693-704.

Long, Z., Zhang, J., Shen, Y., Zhou, C. and Liu, M. 2017. "Polyethyleneimine grafted short halloysite nanotubes for gene delivery." *Materials Science and Engineering C* 81:224-235.

Madhusudana Rao, K., Kumar, A. and Han, S., S. 2018a. "Polysaccharide based hydrogels reinforced with halloysite nanotubes via polyelectrolyte complexation." *Materials Letters* 213:231-235.

Madhusudana Rao, K., Kumar, A., Suneetha, M. and Han, S. S. 2018b. "pH and near-infrared active; chitosan-coated haloysite nanotubes loaded with curcumin-Au hybrid nanoparticles for cancer drug delivery." *International Journal of Biological Macromolecules* 112:119-125.

Maleki, A., Hajizadeh, Z. and Firouzi-Haji, R. 2018. "Eco-friendly functionalization of magnetic halloysite nanotube with SO_3H for synthesis of dihydropyrimidones." *Microporous and Mesoporous Materials* 259:46-53.

Massaro, M., Colletti, C. G., Fiore, B., La Parola, V., Lazzara, G., Guernelli, S., Zaccheroni, N. and Riela, S. 2019. "Gold nanoparticles stabilized by modified halloysite nanotubes for catalytic applications." *Applied Organometallic Chemistry* 33:4665.

Massaro, M., Camofelice, A., Colletti, C. G., Lazzara, G., Noto, R. and Riela, S. 2018a. "Functionalized halloysite nanotubes: Efficient carrier systems for antifungine drugs." *Applied Clay Science* 160:186-192.

Massaro, M., Cavallaro, G., Colletti, C. G., D'Azzo, G., Guernelli, S., Lazzara, G., Pieraccini, S., and Riela, S. 2018b. "Halloysite nanotubes for efficient loading, stabilization and controlled release of insulin." *Journal of Colloid and Interface Science* 524:156-164.

Massaro, M., Riela, S., Baiamonte, C., Blanco, J. L. J., Giordano, C., Lo Meo, P., Milioto, S., Noto, R., Parisi, F., Pizzolanti, G. and Lazzara, G. 2016a. "Dual drug-loaded halloysite hybrid-based glycocluster for sustained release of hydrophobic molecules." *RSC Advances* 6:87935-87944.

Massaro, M., Amorati, R., Cavallaro, G., Guernelli, S., Lazzara, G., Milioto S., Noto, R., Poma, P. and Riela S. 2016b. "Direct chemical grafted curcumin on halloysite nanotubes as dual-responsive prodrug for pharmacological Applications." *Colloids and Surfaces B: Biointerfaces* 140:505-513.

Mehdi, Y. A., Fizir, M., Itatahine, A., He, H. and Dramou, P 2018. "Preparation of multifunctional PEG-graft-Halloysite Nanotubes for Controlled Drug Release, Tumor Cell Targeting, and Bioimaging." *Colloids and Surfaces B: Biointerfaces* 170:322-329.

Michlik, P., Ujhelyiova, A., Krivos, S., Tomcikova, Z. and Hricova, M. 2016. "The structure and properties of polypropylene halloysite nanoclayfibres." *Fibres and Textiles* 23:9-14.

Miller, B. and Keane, C. 2003. *Encyclopedia and Dictionary of Medicine, Nursing, and Allied Health*, 7th Edition. Philadelphia: Saunders.

Mishra, G., and Mukhopadyay, M. 2018. "Enhanced antifouling performance of halloysite nanotubes (HNTs) blended poly(vinyl chloride) (PVC/HNTs) ultrafiltration membranes: For water treatment." *Journal of Industrial and Engineering Chemistry* 63:366-379.

Mousavi, M., Hogsaa, B. and Fini, E. H. 2019 "Intermolecular interactions of bio-modified halloysite nanotube within high impact polystyrene and linear low-density polyethylene." *Applied Surface Science* 473:750-760.

Naumenko, E. A., Guryanov, I. D., Yendluri, R., Lvov, Y. M. and Fakhrullin, R. F. 2016. "Clay nanotube-biopolymer composite scaffolds for tissue engineering." *Nanoscale* 8:7257-7271.

Oliyaei, N., Moosavi-Nasab, M., Tamaddon, A. M. and Fazaeli, M. 2019. "Preparation and characterization of porous starch reinforced with halloysite nanotube by solvent exchange method." *International Journal of Biological Macromolecules* 123:682-690.

Pal, K. 2016a. "Effect of different nanofillers on mechanical and dynamic behavior of PMMA based nanocomposites." *Composites Communications.* 1:25-28.

Pal, P., Kundu, M., K., Maitra, A., Malas, A. and Das, C. K. 2016b. "Synergistic Effect of Halloysite Nanotubes and MA-g-PE on Thermo-Mechanical Properties of Polycarbonate-Cyclic Olefin Copolymer Based Nanocomposite." *Polymer- Plastics Technology and Engineering* 55:1481-1488.

Patel, S., Jammalamadaka, U., Sun, L., Tappa, K. and Mills, D. K. 2016. "Sustained Release of Antibacterial Agents from Doped Halloysite Nanotubes." *Bioengineering* 3:1.

Pavlinakova, V., Fohlerova, Z., Pavlinak, D., Khunova, V. and Vojtova, L. 2018. "Effect of halloysite nanotube structure on physical, chemical, structural and biological properties of elastic polycaprolactone/gelatin nanofibers for wound healing applications." *Materials Science and Engineering* C 91:94-102.

Pietraszek, A., Karewicz, A., Widnic, M., Lachowicz, D., Gajewska, M., Bernasik, A. and Nowakowska, M. 2019. "Halloysite-alkaline phosphatase system-A potential bioactive component of scaffold for bone tissue engineering." *Colloids and Surfaces B: Biointerfaces* 173:1-8.

Pierchala, M. K., Makaremi, M., Tan, H. L., Pushpamalar, J., Muniyandy, S., Solouk, A., Lee, S. M. and Pasbakhsh, P. 2018. "Nanotubes in nanofibers: Antibacterial multilayered polylactic acid/halloysite/ gentamicin membranes for bone regeneration application." *Applied Clay Science* 160:95-105.

Pluta, M., Bojda, J., Piorkowska, E., Murariu, M., Bonnaud, L. and Dubois, P. 2017. "The effect of halloysite nanotubes and N,N'-ethylenebis (stearamide) on the properties of polylactide nanocomposites with amorphous matrix." *Polymer Testing* 61:35-45.

Qiao, X., Na, M., Gao, P. and Sun, K. 2017. "Halloysite nanotubes reinforced ultrahigh molecular weight polyethylene nanocomposite films with different filler concentration and modification." *Polymer Testing* 57:133-140.

Rajan, K. P., Al-Ghamdi, A., Thomas, S. P., Gopanna, A. and Chavali, M. 2017. "Dielectric analysis of polypropylene and polylactic acid blends reinforced with halloysite nanotubes." *Journal of Thermoplastic Composite Materials* 31:1042-1053.

Rinawa, K., Maiti, S. N., Sonnier, R. and Lopez Cuesta, J.-M. 2014. "Influence of microstructure and flexibility of maleated styrene-b-(ethylene-co-butylene)-b-styrene rubber on the mechanical properties of polyamide 12." *Polymer Bulletin* 71:1131-1152.

Risyon, N. P., Othman, S. H., Basha, R. K. and Talib, R. A. 2016. "Effect of Halloysite Concentration and Addition of Glycerol on Mechanical Properties of Bionanocomposite Films." *Polymers & Polymer Composites* 24:795-802.

Sadjadi, S., Malmir, M., Heravi, M. M. and Kahangi F. G. 2018a. "Biocompatible starch halloysite hybrid: An efficient support for immobilizing Pd species and developing a heterogeneous catalyst for ligand and copper free coupling reactions." *International Journal of Biological Macromolecules* 118:1903-1911.

Sadjadi, S., Heravi, M. M. and Kazemi, S. S. 2018b. "Ionic liquid decorated chitosan hybridized with clay: A novel support for immobilizing Pd nanoparticles." *Carbohydrate Polymers* 200:183-190.

Sadjadi, S. and Atai, M. 2018. "Ternary hybrid system of halloysite nanotubes, polyacrylamides and cyclodextrin: an efficient support for immobilization of Pd nanoparticles for catalyzing coupling reaction." *Applied Clay Science* 153:78-89.

Sahnoune, M., Taguet, A., Otazaghine, B., Kaci, M. and Lopez- Cuesta, J.-M. 2017a. "Effects of functionalized halloysite on morphology and properties of polyamide-11/SEBS-g-MA blends." *European Polymer Journal* 90:418-430.

Sahnoune, M., Taguet, A., Otazaghine, B., Kaci, M. and Lopez- Cuesta, J.-M. 2017b. "Inner surface modification of halloysite nanotubes and its influence on morphology and thermal properties of polystyrene/polyamide-11 blends." *Polymer International* 66:300-312.

Sandri, G., Aguzzi, C., Rossi, S., Bonferoni, M. C., Bruni, G., Boselli, C., Cornaglia, A. I., Riva, F., Viseras, C., Caramella, C and Ferrari, F. 2017. "Halloysite and chitosan oligosaccharide nanocomposite for wound healing." *Acta Biomaterialia* 57:216-224.

Sengel, S. B., Sahiner, M., Aktas, N. and Sahiner, N. 2017. „Halloysite-carboxymethyl cellulose cryogel composite from natural sources." *Apllied Clay Science* 140:66-74.

Sharma, N., Ojha, H., Bharadwaj, A., Pal Patak, D. and Sharma, R., Kumar. 2015. "Preparation and catalytic applications of nanomaterials: a review." *RSC Advances* 5:53381-53403.

Shen, X. L., Wu, M. J., Chen, Y. H. and Zhao, G. H. 2010. „Antimicrobial and physical properties of sweet potato starch films incorporated with potassium sorbate or chitosan." *Food Hydrocolloids* 24:285-290.

Shi, R., Niu, Y., Gong, M., Ye, J., Tian, W. and Zhang, L. 2018. "Antimicrobial gelatin-based elastomer nanocomposite membrane loaded with ciprofloxacin and polymyxin B sulfate in halloysite nanotubes for wound dressing." *Materials Science and Engineering C* 87:128-138.

Singh, V. P., Vimal, K. K., Kapur, G. S., Sharma, S. and Choudhary, V. 2016. "High-density polyethylene/halloysite nanocomposites: morphology and rheological behaviour under extensional and shear flow." *Journal of Polymer Research* 23:43.

Smith, R. J., Holder, K. M., Ruiz, S., Hahn, W., Song, Y., Lvov, Y. M. and Grunlan, J. C. 2018. "Environmentally Benign Halloysite Nanotube Multilayer Assembly Significantly Reduces Polyurethane Flammability." *Advanced Functional Materials* 28:1703289.

Spepi, A., Duce, C., Pedone, A., Presti, D., Rivera, J.-G., Ierardi, V. and Tine, M. R. 2016. "Experimental and DFT Characterization of Halloysite Nanotubes Loaded with Salicylic Acid." *The Journal of Physical Chemistry C* 120:26759- 26769.

Sun, J., Yendluri, R., Liu, K., Guo, Y., Lvov, Y. and Yan, X. 2017. "Enzyme-immobilized clay nanotube–chitosan membranes with sustainable biocatalytic activities." *Physical Chemistry Chemical Physics* 19:562-567.

Sun, W., Tang, W., Gu, X., Zhang, S., Sun, J., Li, H. and Liu, X. 2018. "Synergistic effect of kaolinite/halloysite on the flammability and thermostability of polypropylene." *Journal of Applied Polymer Science* 135. DOI: 10.1002/app.46507.

Suppiah, K., Teh, P. L., Husseinsyah, S. and Rahman, R.. 2019. "Properties and characterization of carboymethyl cellulose/halloysite nanotube bio-nanocomposite films: Effect of sodium dodecyl sulfate." *Polymer Bulletin* 76:365-386.

Suppiah, K., Teh P. L., Husseinsyah, S., Rahman, R. and Keat, Y. C. 2017. "Effect of Halloysite Content on Carboxymethyl Cellulose/Halloysite

Nanotube Bio-Nanocomposite Films." *AIP Conferenc Proceedings* 1835:020006-1–020006-5. https://doi.org/10.1063/1.4981828.

Szczepanik, B., Rogala, P., Slomkiewicz, P. M., Banas, D., Kubala- Kukus, A. and Stabrawa, I.. 2017. "Synthesis, characterization, and photocatalytic activity of TiO_2-halloysite and Fe_2O_3-halloysite nanocomposites for photodegradation of chloroanilines in water." *Applied Clay Science* 149:118-126.

Tan, D., Yuan, P., Annabi-Bergaya, F., Dong, F., Liu, D. and He, H. 2015. "A comparative study of tubular halloysite and platy kaolinite as carriers for the loading and release of the herbicide amitrole." *Applied Clay Science* 114:190-196.

Tarasova, E., Naumenko, E., Rozhina, E., Akhatova, F. and Fakhrullin, R. 2019. "Cytocompatibility and uptake of polycations-modified halloysite clay nanotubes." *Applied Clay Science* 169:21-30.

Tas, C. E., Hendessi, S., Baysal, M., Unal, S., Cebeci, F. C., Menceloglu, Y. Z. and Unal, H. 2017. "Halloysite Nanotubes/Polyethylene Nanocomposites for Active Food Packaging Materials with Ethylene Scavenging and Gas Barrier Properties." *Food Bioprocess Technology* 10:789–798.

Teo, Z. X. and Chow W. S. 2016. "Impact, Thermal, and Morphological Properties of PLA- PMMA-Halloysite Nanotube Nanocomposites." *Polymer-Plastics Technology and Engineering* 55, 1474-1480.

Tham, W. L., Poh, B. T., Mohd Ishak, Z. A., Chow, W. S. 2016a. "Transparent Poly(lactic acid)/Halloysite Nanotube Nanocomposites with improved oxygen barrier and antioxidant properties" *Journal of Thermal Analysis and Calorimetry*. 126:1331-1337.

Tham, Wei L., Chow, Wen S., Poh, Beng T., and Mohd Ishak, Zainal A. 2016b. "Poly(lactic acid)/halloysite nanotube nanocomposites with high impact strength and water barrier properties." *Journal of Composite Materials* 50:3925-3934.

Tham, W. L., Poh, B. T., Mohd Ishak, Z. A. and Chow, W. S. 2016c. "Characterisation of Water Absorption of Biodegradable Poly (lactic Acid)/Halloysite Nanotube Nanocomposites at Different Temperatures." *Journal of Engineering Science* 12:13-25.

Tham, W. L., Poh, B. T., Mohd Ishak, Z. A. and Chow, W. S. 2015. "Effect of N,N'-ethylenebis(stearamide) on the water absorption and hydrolytic degradation of poly(lactic acid)/halloysite nanocomposites." *Journal of Thermoplastic Composite Materials*, 30:416- 433.

Therias, S., Murariu, M. and Dubois, P. 2017. "Bionanocomposites based on PLA and halloysite nanotubes: From key properties to photooxidative degradation." *Polymer Degradation and Stability* 145:60-69.

Tian, H., Yan, J., Varada Rajulu, A., Xiang, A. and Luo, X. 2017. "Fabrication and properties of polyvinyl alcohol/starch blend films: Effect of composition and humidity." *International Journal of Biological Macromolecules* 96:518-523.

Tsoufis, T., Katsaros, F., Kooi, B. J., Bletsa, E., Papageorgiou, S., Deligiannakis, Y. and Panagiotopoulos, I. 2017. „Halloysite nanotube-magnetic iron oxide nanoparticle hybrids for the rapid catalytic decomposition of Pentachlorophenol." *Chemical Engineering Journal* 313:466-474.

Tully, J., Yendluri, R. and Lvov, Yuri. 2016. "Halloysite Clay Nanotubes for Enzyme Immobilization." *Biomacromolecules* 17:615-621.

Valapa, R. B., Pugazhenthi, G. and Katiyar, V., 2015. "Effect of graphene content on the properties of poly(lactic acid) nanocomposites." *RSC Advances* 5:28410-28423.

Villmow, T., Pötschke, P., Pegel, S., Häussler, L. and Kretzschmar, B. 2008. "Influence of twin-screw extrusion conditions on the dispersion of multi-walled carbon nanotubes in a poly(lactic acid) matrix." *Polymer* 49:3500-3509.

Vuluga, Z., Corobea, M. C., Elizetxea, C., Ordonez, M., Ghiurea, M., Raditoiu, V., Nicolae, C. A., Florea, D., Iorga, M., Somoghi, R. and Trica, B. 2018. "Morphological and Tribological Properties of PMMA/Halloysite Nanocomposites." *Polymers* 10:816. doi:10.3390/polym10080816.

Wahab, I. F., Abd Razak, S. I. and Yee, F. C. 2017. "Effects of Halloysite Nanotubes on the Mechanical Properties of Polysaccharide Films." *Materials Science Forum* 889:75-78.

Wang, W., Shu, Z., Zhou, J., Li, T., Duan, P., Zhao, Z., Tan, Y., Xie, C. and Cui, S.. 2018. "Halloysite-derived mesoporous g-C_3N_4 nanotubes for improved visible-light photocatalytic hydrogen evolution." *Applied Clay Science* 158:143-149.Wu, D., Li, J., Guan, J., Liu, C., Zhao, X., Zhu, Z., Ma, C., Huo, P., Li, C. and Yan, Y. 2018. "Improved photoelectric performance via fabricated heterojunction g-C_3N_4/TiO_2/HNTs loaded photocatalysts for photodegradation of ciprofloxacin." *Journal of Industrial and Engineering Chemistry* 64:206-218.

Wu, D., Wu, L., Zhang, M. and Wu, L. 2007. „Effect of epoxy resin on rheology of polycarbonate/clay nanocomposites." *European Polymer Journal* 43:1635- 1644.

Wu, W., Cao, X., Luo, J., He, G. and Zhang, Y. 2014. "Morphology, Thermal, and Mechanical Properties of Poly(butylenes succinate) Reinforced with Halloysite Nanotube." *Polymer Composites* 35:847-855.

Wu, F., Zheng, J., Li, Z. and Liu, M. 2019. "Halloysite nanotubes coated 3D printed PLA pattern for guiding human mesenchymal stem cell (hMSCs) orientation." *Chemical Engineering Journal* 359:672-683.

Wu, Y.-P., Yang, J., Gao, H.-Y., Shen, Y., Jiang, L., Zhou, C., Li, Y.- F., He, R.-R. and Liu, M. 2018. "Folate-Conjugated Halloysite Nanotubes, an Efficient Drug Carrier, Deliver Doxorubicin for Targeted Therapy of Breast Cancer." *ACS Applied Nano Materials* 1:595-608.

Xie, A., Dai, J., Chen, X., Zou, T., He, J., Chang, Z., Li, C. and Yan, Y. 2016. "Hollow imprinted polymer nanorods with a tunable shell using halloysite nanotubes as a sacrificial template for selective recognition and separation of chloramphenicol." *RSC Advances* 6:51014-51023.

Yang, J., Wu, Y., Shen, Y., Zhou, C., Li, Y.-F., He, R.-R. and Liu, M.. 2016, "Enhanced Therapeutic Efficacy of Doxorubicin for Breast Cancer Using Chitosan Oligosaccharide-Modified Halloysite Nanotubes." *ACS Applied Materials Interfaces* 8:26578-26590.

Yendluri, R., Otto, D. P., De Villiers, M. M., Vinokurov, V. and Lvov, Y. M. 2017. "Application of halloysite clay nanotubes as a pharmaceutical excipient." *International Journal of Pharmaceutics* 521:267-273.

Yoon, J., T., Jeong, Y., G., Lee, S., C. and Min, B., G. 2009. Influences of poly(lactic acid)-grafted carbon nanotube on thermal, mechanical, and electrical properties of poly(lactic acid). *Polymer for Advanced Technologies* 20:631-638.

Zeng, X., Zhong, B., Jia, Z., Zhang, Q., Chen, Y. and Jia, D. 2019. "Halloysite nanotubes as nanocarriers for plant herbicide and its controlled release in biodegradable polymers composite film." *Applied Clay Science* 171:20-28.

Zeppa, C., Gouanve, F. and Espuche, E. 2009. "Effect of plasticizer on the structure of biodegradable starch/clay nanocomposites: thermal, water-sorption, and oxygen-barrier properties." *Journal of Applied Polymer Science* 112:2044- 2056.

Zhang, J., Luo, X., Wu, Y.-P., Wu, F., Li, Y.-F., He, R.-R. and Liu, M. 2019. "Rod in Tube: A Novel Nanoplatform for Highly Effective Chemo-Photothermal Combination Therapy toward Breast Cancer." *ACS Applid Materials Interfaces* 11:3690- 2703.

Zheng, P., Du, Y., Liu, D. and Ma, X. 2016. "Synthesis, adsorption and photocatalytic property of halloysite-TiO_2-Fe_3O_4 composites." *Desalination and Water Treatment* 57:22703-22710.

Zhong, B., Wang, S., Dong, H., Luo, Y., Jia, Z., Zhou, X., Chen, M., Xie, D. and Jia, D.. 2017. "Halloysite Tubes as Nanocontainers for Herbicide and Its Controlled Release in Biodegradable Poly(vinyl alcohol)/Starch Film." *Journal of Agricultural and Food Chemistry* 65:10445-10451. https://polymerdatabase.com/polymerclasses/Polystyrenetype.html (accessed july 2019).

INDEX

#

2-mercaptobenzothiazole, 139
3-chloropropyl)trimethoxy silane, 145, 149

A

acetic acid, 153
acid, 8, 21, 32, 48, 59, 63, 68, 72, 88, 117, 121, 133, 137, 139, 152, 153, 164, 168, 174
acidic, 8, 69
acrylate, 150
acrylic acid, 140
acrylonitrile, 49, 55
adsorbent, 143
adsorption, 11, 32, 64, 69, 71, 90, 135, 143, 144, 145, 174
Ag/HNT, 143
Ag/HNT/MoS$_2$, 143
agriculture, vii, ix, 57, 65, 138, 155
alkaline phosphatase, 134, 168
allylamine, 129, 132
allyloxy β-cyclodextrin, 131

aluminium, 140
amine, 68, 73, 79, 145, 151, 154
amine-functionalised starch, 145, 154
amino, 88, 138, 150
amino groups, 88, 150
aminoglycoside, 90
ammonia, 10, 15, 19
ammonium, 59, 129
amoxicillin, 132, 133
amplitude, 69
aniline, 151
annihilation, 14
antibiotic, 54, 90, 144
antibiotic residues, 144
anti-corrosion, 42, 49, 50
anticorrosion coating, 12, 14, 139, 155, 161
antifouling, vii, ix, 57, 64, 69, 123, 138, 143, 155, 156, 161, 167
antigen, 163
antioxidant, 68, 81, 132, 171
antitumor, 129, 165
apoptosis, 130
application, 2, 10, 11, 14, 15, 19, 34, 42, 50, 51, 52, 55, 65, 66, 138, 145, 156, 160, 165, 168, 173

aquatic environment, 144
aqueous suspension, 140
aromatic compounds, 120
aromatics, 19
ASP functionalised HNT, 88
aspen-wood-derived bio-crude, 120
atmosphere, 87, 88
atmospheric deposition, 144
atoms, 29, 30, 32
attachment, 18, 19, 25, 29, 50, 157
Au nanoparticles, 34, 69, 152
auramine dye, 144

B

binase-HNT, 129
biocompatibility, 8, 10, 48, 53, 66, 69, 87, 125, 129, 134, 135, 145, 155
biodegradability, 87, 134, 155
biodegradable, 9, 46, 65, 66, 67, 68, 71, 138, 156, 159, 171, 174
biodegradation, 88
biomedical applications, 66, 69, 70
bionanocomposites, 47, 53
bio-PA11, 82
biopolymer, 51, 69, 167
biosensors, 131, 137
blends, 66, 68, 80, 114, 123, 158, 159, 162, 164, 168, 169
bone, 8, 10, 49, 54, 90, 168
BPEI-HNT/PAA-HNT, 140
branched, 113, 117, 140
breast cancer, 20, 130
brilliant green, 62, 129, 132, 133
brittleness, 48, 87
bulk polymerization, 44, 53
butadiene, 49, 55
butadiene-styrene, 49

C

caffeic acid, 68, 72
calorimetry, 53, 55
cancer, 69, 124, 129, 130, 135, 155, 166
cancer cells, 69, 124, 129, 130
cancer therapy, 125, 130, 132, 133, 155, 162
candidates, viii, 18, 41, 44, 134
carbon, 2, 12, 32, 43, 48, 50, 87, 109, 126, 143, 151, 163, 172, 174
carbon dioxide, 151
carbon monoxide, 32
carbon nanotubes, 2, 12, 109, 163, 172
carbonyl groups, 88
carboxyl, 163
carboxylic acid, 89
carboxymethyl cellulose, 169
carcinoma, 129
cardanol, 132
carrageenan, 59, 71, 78
casting, 47, 72, 73, 74, 77, 90, 104, 106, 107, 121, 126, 134
catalysis, vii, viii, ix, 18, 29, 33, 57, 145, 151, 154, 155
catalyst, viii, 2, 10, 145, 149, 150, 151, 152, 156, 158, 169
catalytic activity, 19, 33, 146, 149, 151, 153, 158
catalytic effect, 129
catalytic properties, 19, 163
cell culture, 137
cell line, 129
cellulose, 59, 65, 71, 158, 169, 170
ceramic, vii, viii, 2, 9
challenges, 42
chemical, vii, 1, 2, 3, 8, 18, 20, 26, 28, 50, 53, 123, 124, 129, 131, 137, 143, 145, 152, 155, 156, 166, 168
chemical inertness, 123, 124
chemical properties, 2, 50
chemical stability, 28

chitosan, 13, 46, 47, 49, 52, 53, 54, 65, 66, 68, 69, 75, 129, 130, 131, 132, 134, 135, 136, 145, 149, 154, 157, 160, 163, 165, 166, 169, 170, 173
chitosan-agarose-gelatine, 134
chitosan-alginate, 134
chloramphenicol, 144, 173
chlorhexidine, 132, 133, 161
chloroperoxidase, 145, 161
ciprofloxacin, 135, 136, 170, 173
clay mineral, v, vii, 1, 2, 3, 8, 10, 12, 13, 14, 15, 33, 51, 52, 79, 140, 160
clay minerals, 8, 52, 79, 140
composites, v, vii, viii, ix, 2, 11, 17, 19, 28, 33, 34, 37, 42, 52, 53, 54, 55, 57, 64, 65, 66, 67, 68, 72, 79, 81, 82, 83, 84, 85, 86, 87, 91, 92, 109, 110, 111, 112, 115, 120, 121, 122, 123, 135, 138, 139, 140, 155, 156, 158, 160, 161, 163, 164, 167, 169, 173, 174
compounds, 52, 120, 124, 129, 130, 132, 140
copolymer, 14, 54, 59, 61, 66, 80, 121, 123, 124, 126, 138
copolymerisation, 120
copper, 11, 25, 69, 139, 146, 150, 156, 157, 160, 169
corrosion, 9, 42, 48, 50, 54, 55, 139
cost, ix, 2, 43, 50, 57, 65, 139, 155
covalent bond, 149
covalent bonding, 149
crystal structure, 3, 4, 81
crystalline, ix, 57, 69, 88
crystallinity, 62, 65, 67, 82, 88, 89, 90, 92, 114, 115, 117, 119
crystallisation, 66, 67, 80, 82, 88, 89, 92, 114
crystallization, vii, viii, 1, 2, 41, 44, 162, 164, 165
crystals, 7, 29, 30, 31, 32, 114
curcumin, 69, 75, 130, 131, 132, 133, 165, 166

cyclodextrin, 20, 131, 133, 150, 154, 169

D

data analysis, 64
database, 120
decomposition, 5, 10, 15, 19, 29, 44, 69, 140, 151, 153, 172
decomposition temperature, 44
degradation, 19, 82, 88, 145, 161, 165, 172
degradation process, 82, 88
degree of crystallinity, 44, 66, 125
dehydrated halloysite, 3, 4, 6
dendrimer decorated HNT, 150
dextrin, 129, 133
diatomite, 81, 83, 115, 118, 160
diffusion, 67, 87
dihydropyrimidinones, 152
dispersion, 10, 42, 44, 47, 48, 50, 53, 67, 71, 79, 109, 113, 114, 120, 130, 138, 142, 155, 159, 172
distribution, 22, 23, 25, 26, 27, 49, 108, 123
doxorubicin, 20, 62, 129, 132, 133, 163, 165, 173
doxycycline, 132, 133
drainage, 144
drawing, 82, 85, 119, 130
drinking water, 143
drug career, 19
drug delivery, viii, 18, 19, 42, 43, 48, 50, 55, 69, 70, 124, 131, 155, 158, 159, 161, 165, 166
drug release, 49, 55, 65, 90, 103, 133
drugs, vii, viii, 2, 19, 64, 66, 132, 137, 156, 166
dynamic mechanical analysis, 48
dynamic-mechanical properties, 164

E

electrical properties, 174

electrospinning, 48, 70, 90
elongation, 9, 66, 72, 73, 74, 76, 77, 78, 81, 83, 84, 85, 86, 90, 92, 94, 95, 96, 98, 99, 102, 103, 104, 107, 111, 113, 116, 117, 118, 119, 121, 126, 127, 141
emulsion polymerization, 45, 53
encapsulation, vii, viii, 2, 29, 55, 69, 90, 103, 135, 140, 156
engineering, 49, 51, 54, 66, 123, 124, 134
entanglements, 80
entrapment, 113, 114
environment, 8, 64, 65, 138, 144, 156
environment protection, 64, 141
epoxy, 47, 53, 139, 173
ethanol, 20, 22, 26, 29, 68, 71, 130
ethylene, 9, 13, 14, 55, 59, 119, 124, 132, 133, 138, 159, 168
ethylene glycol, 132, 133
EVA-TCPM-halloysite, 138
experimental condition, 21, 24, 26
experimental design, 30, 134
extrusion, 43, 49, 68, 69, 108, 160, 172

F

fibroblasts, 15, 82, 135
fillers, 65, 66, 79, 87, 92, 109, 113, 114, 125, 140, 160
films, 10, 15, 47, 68, 71, 89, 90, 92, 114, 115, 157, 159, 168, 170
fire resistance, 67, 115
fire retardancy, 49, 140, 141, 156
flame, 9, 13, 48, 59, 64, 79, 81, 113, 114, 140, 155, 156, 159, 162, 163, 165
flame retardant, 49, 59, 64, 81, 114, 140, 142, 155, 156, 162, 163
flammability, vii, viii, 41, 42, 49, 50, 55, 82, 115, 124, 140, 170
flexibility, 66, 71, 168
fluid, 7, 69, 144
fluorescein isothiocyanate, 63, 129

fluorescence, 131, 132
folate, 129
folic acid, 63, 129
food, vii, ix, 57, 87, 90, 120, 138, 139, 155, 156
food packaging, vii, ix, 57, 87, 115, 120, 139, 155, 156, 171
formation, viii, 7, 18, 19, 20, 21, 22, 24, 25, 26, 27, 28, 29, 30, 31, 32, 53, 67, 89, 114, 135
formula, vii, 1, 3, 18, 82
functionalised HNT, 80, 130, 149, 150, 151
functionalization, 29, 50, 166

G

gelatine, 84, 134, 135, 136
gene therapy, 131
gentamicin, 103, 132, 168
geology, 15
glass transition, 44, 48
glass transition temperature, 44, 48
glucose, 65, 131, 133
glucose oxidase, 131, 133
glycerol, 71, 77, 92, 106, 135, 136, 169
green nanocomposites, 49
growth, vii, viii, 17, 18, 20, 23, 29, 30, 31, 32, 70, 114, 150
growth rate, viii, 18, 23, 30, 31, 32, 70

H

halloysite, v, vii, viii, ix, 1, 2, 3, 4, 5, 6, 7, 8, 9, 10, 11, 12, 13, 14, 15, 17, 18, 33, 34, 35, 36, 37, 41, 42, 43, 45, 47, 48, 50, 51, 52, 53, 54, 55, 56, 57, 58, 59, 64, 65, 79, 87, 125, 137, 138, 139, 140, 141, 143, 145, 150, 157, 158, 159, 160, 161,162, 163, 164, 165, 166, 167, 168, 169, 170, 171, 172, 173, 174
hardness, 66, 68, 109, 114

hazardous substance, 138
hazardous substances, 138
HDPE/HDPE-g-MA/HNT, 113
heavy metal ions, 65, 141, 144, 156, 163
heavy metals, 144
herbicide, 64, 138, 171, 174
herbicide amitrole, 138, 171
heterogeneous catalysis, 151
heterogeneous catalyst, 149, 150, 151, 157, 158, 169
high strength, 88
histogram, 23, 25, 27
HNT alginate, 135
HNT/chitosan, 69, 74, 75, 134, 135, 136
HNT/CMC, 68, 73
HNT/enzyme chitosan, 69
HNT/insulin, 131
HNT/PBAT, 67, 72
HNT/PTFE, 124, 126, 127, 128
HNT/UHMWPE, 114, 117
HNTs/PE, 115
HNTs-g-PEG-CDs-Biotin, 125
HNTs-PVA, 67
hybrid, 9, 55, 69, 92, 129, 130, 131, 145, 149, 150, 153, 166, 169
hydrated halloysite, 3, 8
hydrogels, 68, 69, 134, 163, 166
hydrogen, 10, 15, 32, 47, 145, 173
hydrogen peroxide, 32
hydrogenation, 10, 19, 153

I

ibuprofen, 9, 14, 134
imine-functionalisation, 151
immobilization, 161, 169
immunoglobulin, 132
impact strength, 53, 66, 85, 87, 88, 93, 94, 95, 98, 99, 110, 111, 112, 113, 114, 115, 116, 117, 118, 120, 125, 127, 128, 141, 171

in vitro, 15, 129, 130, 131, 134, 137
in vivo, 12, 20, 129, 134, 135
industry, vii, viii, 2, 9, 66, 108, 120, 123
in-situ growth, v, vii, viii, 17, 18, 20, 29
insulin, 131, 133, 166
interface, 53, 67, 80, 109, 123, 155
interfacial adhesion, 71, 113
interfacial bonding, 109
interphase, 120
intrinsic viscosity, 88
iodine, 132, 133
ions, 21, 26, 28, 141, 144
IR spectroscopy, 13
iron, 7, 8, 91, 152, 153, 172
irradiation, 20, 145, 149, 150, 165

K

kaolinite, 2, 3, 6, 7, 12, 43, 115, 170, 171
kaolinite/HNT, 115
KH550, 60, 114, 115, 117, 136
kinetics, 30, 89

L

laccase, 131
lactic acid, 58, 159, 161, 171, 172, 174
ligand, 11, 146, 150, 157, 169
light, 50, 81, 88, 151, 165, 173
light transmittance, 88
limestone, 8
lipase, 69, 70, 75, 131, 133
liquids, 149
lithium, 68
LLDPE, 9, 13, 58, 119, 120, 159
low density polyethylene, 9
low temperatures, 79
lumen, 7, 20, 23, 43, 44, 69, 70, 81, 82, 84, 129, 130, 135, 137, 138, 139, 140, 143, 155, 156
L-Valine, 139, 161

lysozyme, 69, 70, 75, 90, 102, 103, 159

M

macromolecules, 53
magnesium, 140
magnetite nanoparticles, 153
MA-g-PE, 124, 167
MA-g-PHBV, 66
manufacturing, viii, 41
marine environment, 138
mass, 43, 117, 140
materials, viii, 9, 10, 18, 41, 42, 43, 47, 48, 49, 51, 53, 54, 55, 61, 65, 66, 109, 115, 135, 137, 153, 157, 162
matrix, 43, 47, 48, 71, 79, 80, 81, 89, 90, 91, 92, 108, 113, 114, 115, 117, 120, 123, 168, 172
mechanical performances, 125
mechanical properties, 8, 9, 44, 47, 48, 49, 51, 52, 53, 54, 55, 65, 66, 67, 68, 69, 70, 71, 80, 81, 82, 87, 88, 90, 92, 109, 114, 115, 123, 124, 134, 135, 141, 155, 157, 159, 160, 161, 162, 168
media, 26, 29, 31, 32, 144, 150, 151
medical, 10, 48, 65, 66, 71, 82, 120, 123, 162
melt, 42, 43, 49, 66, 82, 89, 90, 91, 109, 113, 116, 141
melting, 82, 87
melting temperature, 87
membrane, 67, 69, 70, 90, 122, 123, 143, 170
membranes, 42, 69, 123, 124, 143, 159, 167, 168, 170
metal ion, 29, 30, 32, 65, 141, 144, 156, 163
metal nanoparticles, vii, 163
metal phosphinate, 59, 85, 141, 142
metal salts, 153
metal-based catalysts, 145
methyl methacrylate, 54, 159

microhardness, 88, 111
microorganisms, 138, 143
microspheres, 45, 53, 69, 165
microstructure, 69, 114, 168
mixing, 75, 91, 108, 113, 134, 136, 138
modification agents, 68
modulus, 9, 44, 47, 48, 66, 70, 71, 72, 73, 74, 76, 77, 78, 81, 82, 83, 84, 85, 86, 88, 89, 91, 92, 94, 95, 96, 97, 98, 99, 101, 102, 103, 104, 105, 107, 109, 110, 111, 113, 114, 116, 117, 118, 119, 121, 123, 125, 126, 127, 128, 141
moisture content, 73
molecular weight, 53, 58, 163, 168
molecules, 3, 5, 11, 43, 64, 65, 67, 82, 135, 166
morphology, ix, 6, 7, 23, 29, 31, 50, 51, 57, 68, 69, 80, 81, 88, 91, 92, 113, 114, 123, 134, 162, 169, 170, 173
moulding, 108, 126, 127, 128

N

nanocomposites, viii, 9, 13, 18, 19, 20, 22, 26, 27, 28, 29, 41, 43, 44, 48, 49, 50, 51, 52, 53, 54, 55, 64, 67, 68, 69, 70, 71, 72, 81, 82, 83, 87, 88, 89, 90, 91, 92, 93, 108, 109, 110, 113, 114, 115, 116, 120, 121, 123, 124, 125, 126, 132, 135, 141, 142, 151, 155, 157, 158, 159, 162, 163, 164, 167, 168, 170, 171, 172, 173, 174
nanocontainers, 12, 14, 64, 82, 137, 139, 156, 174
nanocrystals, 30, 31, 71
nanofibers, 48, 54, 70, 135, 160, 168
nanomaterials, viii, 17, 51, 169
nanomedicine, 54
nanoparticles, vii, viii, 10, 15, 17, 18, 34, 35, 36, 37, 38, 39, 50, 69, 92, 125, 150, 152, 153, 155, 156, 161, 163, 166, 169
nanopeapods, 22, 37

nanoreactors, viii, 18
nanorods, viii, 18, 143, 144, 173
nanostructured materials, 50
nanostructures, viii, 18, 19, 20, 23, 24, 28, 29, 30, 32, 42, 50
nanotechnology, 108, 138
nanotubes, v, vii, viii, ix, 2, 9, 10, 11, 12, 13, 14, 15, 17, 18, 19, 33, 34, 35, 36, 38, 41, 44, 46, 47, 48, 50, 51, 52, 53, 54, 55, 56, 57, 59, 64, 65, 79, 109, 114, 123, 125, 134, 137, 138, 139, 140, 141, 143, 145, 156, 157, 158, 159, 160, 161, 162, 163, 164, 165, 166, 167, 168, 169, 170, 171, 172, 173, 174
nitrites, 143
non-flammability, vii, viii, 41, 42, 49, 50
non-polar, 120
nontoxicity, 66, 139
norfloxacin, 130, 133
nucleating agent, 80, 92, 123
nucleation, viii, 12, 18, 21, 23, 29, 30, 31, 32, 67, 88, 89

O

occurence, 2
oleic acid, 20, 114
oligosaccharide, 129, 132, 169
organic, 19, 29, 39, 52, 64, 109, 123, 144, 145, 151, 156
organic compounds, 151
organic solvents, 123, 145
organophosphorus, 81, 140
oxidase, 131, 133
oxidation, 19, 143
oxidative reaction, 82
oxygen, 87, 88, 97, 98, 115, 120, 125, 127, 171, 174

P

PA 11, 43, 79
PA 6, 79
PA11, ix, 57, 58, 79, 80, 82, 83, 85, 86, 121, 123, 140, 142, 155, 156
PA11/HNT, 82
PA6/HNT, 81, 83, 84
PA6/NBR, 80, 83
PBAT, 58, 66, 67, 162, 165
PBS, 58, 65, 66, 72, 164
PC, 58, 123, 124, 126
PC/COC, 124, 126
PCL/Gelatine, 82, 136
PCL/Gelatine/HNT, 82
PCL/PLA, 82, 85
Pd, viii, 10, 18, 19, 20, 28, 29, 30, 32, 145, 149, 150, 153, 154, 156, 158, 169
Pd nanoparticles, 150, 156, 169
Pd-based halloysite –chitosan-sulfur-functionalised ionic-liquids, 149
PE-g-MA, 59, 114, 116, 126
PEG-NH2, 124, 126
PEI-modified HNTs, 114
permeability, 77, 88, 97, 98, 108, 127
pH, 19, 50, 69, 100, 101, 131, 139, 145, 152, 156, 166
pharmaceutical, vii, viii, 2, 9, 11, 15, 66, 131, 135, 155, 164, 173
pharmacotherapy, vii, ix, 57, 155
PHB/PCL, 68, 72
PHBV, 58, 65, 66, 72, 158, 164
PHBV/PBS blends, 66
phosphate, 59, 60, 140
phosphorus, 165
photocatalysts, 151, 157, 165, 173
photodegradation, 171, 173
photooxidation, 91
photooxidative degradation, 91, 172
physical properties, 2, 42, 66, 123, 159, 170

PLA, ix, 58, 82, 87, 88, 89, 90, 91, 92, 93, 94, 95, 96, 97, 98, 99, 100, 101, 102, 103, 104, 105, 106, 107, 112, 114, 117, 136, 137, 140, 141, 142, 155, 156, 158, 159, 162, 171, 172, 173
PLA/ASP-HNTs, 88
PLA/HNT, 87, 88, 89, 90, 91, 92, 158
PLA/HNTs- ZrO_2, 92, 158
PLA/PMMA/HNT, 89
PMMA/HNT-EBS, 108
polar, viii, 18, 29, 120, 124
polarity, 109, 124
politics, 13
pollutants, 19, 29, 141, 144, 151, 156
pollution, 65, 140, 141
poly (allylamine), 129
poly(acrylic acid), 140
poly(diallyldimethyl-ammonium), 129
poly(ethylenimine), 129
poly(methyl methacrylate), 42, 48
poly(vinyl chloride), 167
polyacrylamide, 150
polyamide, 42, 43, 52, 58, 79, 82, 141, 159, 163, 168, 169
polycaprolactone/gelatine, 135
polycarbonate, 173
polyelectrolyte complex, 69, 166
polyethylenimine, 73, 140
polylactic acid, 87, 137, 160, 164, 168
polymer, vii, viii, ix, 1, 9, 10, 11, 13, 14, 34, 41, 42, 43, 46, 47, 48, 50, 51, 52, 53, 54, 55, 57, 61, 64, 65, 66, 71, 72, 79, 80, 81, 82, 87, 88, 92, 108, 109, 113, 115, 120, 123, 124, 130, 131, 140, 142, 143, 145, 155, 158, 159, 160, 161, 162, 163, 164, 165, 167, 168, 169, 170, 171, 172, 173, 174
polymer blends, 159
polymer composites, vii, viii, ix, 2, 11, 42, 57, 64, 140, 155
polymer materials, 155

polymer matrix, ix, 50, 57, 79, 80, 109, 123, 124, 155
polymer molecule, 92
polymer nanocomposites, 51, 55, 71, 109
polymeric materials, 140
polymerization, 45, 53, 144
polymers, 9, 42, 49, 50, 53, 59, 65, 79, 87, 109, 113, 124, 138, 140, 143, 155, 156, 174
polymethylmethacrylate, 92
polymyxin B sulphate, 135, 136
polyolefin, 109, 116, 120
polyolefins, 87, 109
polypropylene, 163, 167, 168, 170
polysaccharide, 46
polysaccharides, 59, 65
polystyrene, 14, 42, 44, 45, 53, 58, 120, 130, 158, 159, 167, 169
polysulfone, 143
polyurethane, 42, 49, 140
polyurethane foam, 140
polyurethanes, 49
polyvinyl alcohol, 172
polyvinylidene fluoride, 143
porosity, 43, 71, 134
potassium, 12, 68, 132, 150, 170
potassium clavulanate, 132
PP/ABS, 114, 117, 164
PP/HNT, 115, 116, 117, 118
precipitation, 19, 144
preparation, iv, 23, 42, 54, 68, 70, 90, 125, 152, 160
proliferation, 134, 137, 138
propylene, 9, 13, 14, 55
protection, 9, 11, 48, 64, 90, 138
proteins, 131, 133
PS/PA11, 80, 121, 123
Pt, 153
purification, 22, 143, 159
PVA/ST, 138
PVA/starch, 71
PVC/HNTs membrane, 143

Index

R

reaction time, 21, 24, 26, 29, 32, 151
reactions, 19, 145, 146, 147, 148, 149, 150, 151, 152, 157, 158, 169
redundancy, 65, 151
regeneration, 43, 90, 144, 168
reinforcement, ix, 57, 64, 71, 79, 81, 90, 155
removing, 144
researchers, 139, 143, 156
resistance, 48, 49, 66, 79, 86, 92, 108, 109, 113, 115, 120, 123, 124, 155
Rh, 153
rheology, 43, 91, 173

S

salicylic acid, 139, 170
scanning calorimetry, 63
scanning electron microscopy, 7
science, 36, 38, 51, 53
SEBS-g-MA, 59, 80, 83, 87, 93, 94, 117, 160, 169
silane, 11, 60, 72, 80, 88, 135, 136, 145, 149, 164
silibin, 131
silica, 7, 9, 11, 66
silver, 19, 26, 143, 162
sintering, 126, 127, 128
sodium, 11, 15, 32, 45, 53, 151, 170
sodium dodecyl sulfate, 45, 53, 170
sodium dodecyl sulfate (SDS), 45
sodium polystyrene sulphonate, 130
solution, 21, 22, 30, 32, 42, 47, 68, 90, 141, 143, 144, 152, 153, 156
solvents, viii, 18, 20, 22, 23, 26, 27, 29, 30, 31, 32
specific surface, viii, 2, 5, 11, 43
stability, 14, 26, 29, 33, 49, 66, 69, 92, 109, 120, 123, 130, 145, 161

stabilization, 166
starch, ix, 58, 65, 71, 77, 78, 138, 145, 154, 155, 157, 167, 169, 170, 172, 174
storage, 65, 83, 88, 89, 91, 101, 104, 105, 117, 125, 137, 139, 156
stress, 66, 80, 83, 124, 126, 134
structure, vii, ix, 1, 2, 3, 6, 10, 19, 24, 42, 43, 53, 57, 66, 70, 81, 109, 115, 131, 134, 139, 140, 145, 155, 156, 167, 168, 174
styrene, 10, 44, 45, 55, 59, 121, 158, 168
styrene polymerization, 45
sulphate, 12, 103, 132
superparamagnetic, 152, 157
superparamagnetic heterogeneous nanocatalyst, 152, 157
support, viii, 2, 10, 15, 145, 150, 156, 169
surface area, viii, 5, 9, 41, 42, 145, 156
surface chemistry, 32
surface modification, 50, 123, 164, 169
surface properties, 144
surface tension, 30
surface treatment, 92
surfactant, 30, 31, 32, 46, 53, 62, 68
sweet potato starch, 170
swelling, 12, 43, 66, 69, 114
synergistic effect, 55, 71, 81, 108, 114, 115, 124
synthesis, 9, 15, 20, 26, 29, 31, 32, 34, 35, 36, 37, 38, 39, 54, 70, 152, 153, 163, 166, 171, 174

T

techniques, 6, 20, 42, 50, 67, 143, 144
temperature, vii, viii, 3, 4, 5, 17, 20, 21, 29, 30, 32, 44, 82, 89, 90, 109, 124, 131, 144
tensile strength, 9, 44, 47, 70, 71, 72, 73, 74, 76, 77, 78, 81, 82, 83, 84, 85, 86, 88, 91, 92, 94, 95, 96, 98, 102, 103, 104, 105, 106, 107, 108, 110, 111, 113, 114, 116,

117, 118, 119, 120, 121, 122, 123, 124, 125, 126, 127, 128, 141
therapy, 131, 132, 133
thermal analysis, 53, 55
thermal behaviour, 5, 67
thermal decomposition, 140
thermal properties, 66, 71, 80, 81, 87, 88, 89, 90, 91, 114, 123, 158, 162, 164, 169
thermal stability, vii, viii, ix, 9, 41, 43, 44, 48, 49, 50, 57, 66, 67, 68, 80, 81, 87, 88, 89, 92, 114, 115, 123, 124, 125, 131, 134, 140, 141, 144, 145, 155
thermal treatment, 3
thermodynamics, 30
thermogravimetric analysis, 5
thermooxidative stability, 88
thermoplastics, 156
thymol, 82, 85
tissue, 10, 13, 42, 47, 48, 50, 52, 56, 66, 70, 134, 155, 167, 168
tissue engineering, 10, 13, 42, 47, 48, 50, 52, 56, 66, 134, 155, 167, 168
toluene, 20, 22, 26, 29, 149
toxicity, 19, 125, 129, 138, 143, 155
transparency, 43, 48, 51, 79, 87, 88
treatment, vii, viii, ix, 3, 14, 18, 19, 20, 37, 48, 52, 57, 67, 68, 79, 81, 88, 92, 113, 114, 115, 123, 124, 125, 135, 141, 144, 145, 149, 155, 156, 167, 174
triazolium salts, 132
trichlorophenyl maleimide, 138

V

vacuum, 135, 137, 153, 156
vancomycin, 135, 136, 164
variations, 23, 72, 73, 74, 75, 76, 77, 78, 83, 84, 85, 86, 93, 94, 95, 96, 97, 98, 99, 100, 101, 102, 103, 104, 105, 106, 107, 109, 110, 111, 112, 116, 117, 118, 119, 121, 122, 126, 127, 128

W

wastewater, 123, 141, 143, 144, 145, 156
water, vii, viii, ix, 3, 5, 18, 29, 33, 52, 57, 66, 67, 68, 71, 77, 87, 88, 89, 90, 98, 99, 100, 101, 102, 103, 115, 123, 125, 126, 127, 134, 139, 141, 143, 151, 153, 155, 159, 167, 171, 172, 174
water absorption, 68, 87, 88, 89, 172
water diffusion, 88, 89
water purification, 159
water resources, 141
water sorption, 71
water vapor, 102, 103
wound healing, 135, 136, 155, 168, 169

X

xanthan gum, 61, 69, 75

Z

ZrO_2 nanoparticles, 92

Related Nova Publications

Recent Advances in Geophysics

Editor: Christina N. Brandt

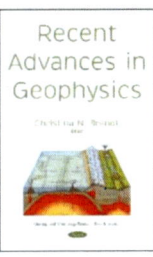

Series: Geology and Mineralogy Research Developments

Book Description: In the opening chapter of this compilation, the authors process and geologically interpretate the marine geological mapping of a detailed grid of very high resolution seismic profiles recorded in the Campania continental shelf between the Solofrone river mouth and Agnone.

Softcover ISBN: 978-1-53616-207-3
Retail Price: $82

Fluorite: Structure, Chemistry and Applications

Editor: Moritz van Asten

Series: Geology and Mineralogy Research Developments

Book Description: In *Fluorite: Structure, Chemistry and Applications,* crystal structure of fluorites, point defects, specific properties, innovative methods of nanopowders synthesis and properties of ceria-based materials are discussed.

Softcover ISBN: 978-1-53615-204-3
Retail Price: $95

To see a complete list of Nova publications, please visit our website at www.novapublishers.com

Related Nova Publications

FAULT-CONTROLLED PROCESSES OF BASIN EVOLUTION: A CASE ON A LONGSTANDING TECTONIC LINE

AUTHORS: Yasuto Itoh and Mitsuru Inaba

SERIES: Geology and Mineralogy Research Developments

BOOK DESCRIPTION: The book attempts to unravel the causal relationship between transient modes of subduction of the oceanic plate and considerable diversity of basinal morphologies developed upon the convergent margin.

EBOOK ISBN: 978-1-53614-676-9
RETAIL PRICE: $82

SEDIMENTARY BASINS: EVOLUTION, METHODS OF FORMATION AND RECENT ADVANCES

EDITOR: Sam Brookes

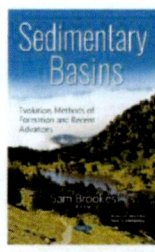

SERIES: Geology and Mineralogy Research Developments

BOOK DESCRIPTION: In this collection, the sedimentary basins of the northern Campania Tyrrhenian margin have been investigated in detail aimed at studying and reconstructing their Quaternary geologic evolution through seismo-stratigraphic data.

SOFTCOVER ISBN: 978-1-53613-922-8
RETAIL PRICE: $82

To see a complete list of Nova publications, please visit our website at www.novapublishers.com